Making Ammonia

Benjamin Johnson

Making Ammonia

Fritz Haber, Walther Nernst, and the Nature
of Scientific Discovery

 Springer

Benjamin Johnson
Max Planck Institute for the History
of Science
Berlin, Germany

Max Planck Institute for Chemical
Energy Conversion
Mülheim, Germany

ISBN 978-3-030-85534-5   ISBN 978-3-030-85532-1   (eBook)
https://doi.org/10.1007/978-3-030-85532-1

The discoverer always has an "extraordinary" background—either special knowledge which prepares him uniquely to "infer" his discovery, or special ignorance which allows him, oblivious to the directives of orthodox doctrine, to stumble upon his discovery.

—Paul Forman, 1970

If my calculations are correct, when this baby hits 88 miles per hour, you're gonna see some serious shit.

—Doc Brown, 1985

For Everyone

Preface by Gerhard Ertl

According to the will of Alfred Nobel, his prize should be awarded to a person who made the contribution "for the greatest benefit of mankind." From this point of view, hardly anybody else deserved this distinction more than Fritz Haber for his successful experiment in July 1909. The history of this event and its further technical development have already been extensively described in the literature, but the present work goes far beyond this approach. In particular, it emphasizes the importance of interactions between scientists leading to the exchange of knowledge. For example, Haber's success would not have been possible without his controversy with Walther Nernst. (Ironically, Nernst received the big Prize just one year after Haber.) As a whole, this development is embedded in the period of growth of thermodynamics and physical chemistry.

After analyzing the confluence of factors that lead to a scientific breakthrough, the author makes in the last part an attempt to formulate a theory for the occurrence of a breakthrough in science which he calls the *Haze*, starting from Kuhn's concept of a paradigm shift. Thus, this text will be of interest to anybody working on the progress of science.

Gerhard Ertl

Preface

This book is about the natural sciences.

It is an account of a unique discovery in the history of science: the synthetic production of ammonia from the elements, known today as the Haber-Bosch process. The story illustrates how we are able to change our lives through scientific investigation to make possible the previously inconceivable.

Despite our accomplishments, however, we still have difficulty answering basic questions. What is science? And how does science work? It is certainly more than an exclusive engagement with hard facts and truths, as is often posited. At its core, science may be characterized as the interpretation of experimental observations within a theoretical framework (although preliminary planning and equipment are also required). Fittingly or paradoxically, the way in which science works cannot be described so succinctly. My attempt to formulate an answer fills the pages of this book. And as the pages demonstrate, performing scientific research is not an exact science (hah!). Scientific research is a trade. Science thrives on the accumulated knowledge of past generations and science exists because we practice it.

That's right, we *practice* science, because science is an active undertaking.

Today, science is often regarded as something like a religion. Unquestioning allegiance is expected in the face of definitive scientific results. *Do you believe in science?* I don't, not in any faith-based sense. But I do believe that scientific research provides us with valuable information to help shape our world in the way we see fit. Decision makers often ask, are we following the science? Do we have the science to make an informed decision? As I see it, what we *have* is evidence. We have numbers, measurements, and observations. When we practice science, we make interpretations based on this evidence in order to inform our decision-making. It is not a process that stops because we have found *The Answer*.

Now, an answer to a scientific question does exist, at least in terms of the theories (paradigms) we have developed. In actively practicing science, we come ever closer to these answers in an informed, inquisitive, and critical way. We constantly refine results and ask what the results mean in different contexts. Dissent is welcome and necessary. Controversy and the discussion of objections exemplify the interaction between scientists, which leads to deeper insight.

How does this process work exactly? The best way to answer is to show you, not to tell you. The detailed case study in this book, the famous and consequential scientific breakthrough of ammonia synthesis from the elements, is a great place to start. The historical record is rich, and with hindsight, we can understand the scientific developments well. What will the story reveal to you?

The book is divided into three parts. Part I, A Confluence of Factors, summarizes a century of scientific advances leading to the arena for discovery in which the breakthrough of ammonia synthesis took place. The developments of this discovery between 1903 and 1909 are given in detail in Part II, The Scientific Breakthrough. The events are examined within the context of physical chemistry and the relationship between Fritz Haber and Walther Nernst, the physical chemists at the center of research on ammonia synthesis. Part III, The Haze, is a theoretical examination, based on interactions between scientists. In it, I generalize the events and dynamics leading to a scientific breakthrough as described in the first two parts.

The narrative is indeed a celebration of the successes of science. I wrote this book from the perspective of a scientist who holds scientific research in high regard because of its potential. I do, of course, also have criticisms of how we practice science today and concede there can be unintended and adverse consequences to scientific and technological developments. Some such consequences are addressed in the book. The story of ammonia synthesis illustrates well the challenges connected with the scientific endeavor.

Another point of discussion is the status of a promising mode of research still stuck in its infancy despite its long history. This is the interdisciplinary approach. Cross-disciplinary research is stagnant because it is risky, and it is risky because it is difficult. However, risk taking should not be avoided. Progress on ammonia synthesis was repeatedly dependent on bringing together—with uncertain consequences—previously unconnected pieces of information from different fields. The potential reward of this risk-taking is apparent throughout Parts I and II of this book. In fact, the existence of the book itself is the result of an expedition into an uncharted mixture of experiences, expertise, and expectations. Looking back to the beginning, the outcome was by no means assured.

In this spirit, the book is not only an effort to convey knowledge and offer useful answers about how we practice science, it is an attempt to show the results of taking the leap. Like an experiment, some things are worth doing just to see what happens.

B.E.J.
Berlin, Germany
June 2022

Acknowledgments

The completion of this book was dependent on many people to whom I am grateful and indebted.

Thank you to:

My friends and family for supporting me no matter what I get myself into.

Dr. Kristina Starkloff, Susanne Uebele, Florian Spillert, Bernd Hoffmann, and all at the Archive of the Max Planck Society

The staff at the Staatsbibliothek zu Berlin—especially the Unter den Linden branch

All at the Max Planck Institute for the History of Science (MPIWG) and the Fritz Haber Institute of the Max Plank Society (FHI) for enabling me to dedicate myself to this project for 5 years.

Two anonymous referees for the global and specific suggestions that helped more clearly define the objectives of the book.

Dr. Travis Jones at FHI for the discussions on theoretical thermodynamics, physical chemistry, and the nature of science in general.

Urte Brauckmann at MPIWG for helping me navigate the world of image copyrights.

Prof. David Obstfeld at California State University, Fullerton for guidance on how to employ social network theory in the context of science.

Dipl.-Psych. Claudia Walther at the Freie Universität Berlin for the suggestion and subsequent discussions of social network theory.

Prof. Kathryn Olesko at Georgetown University for showing me the ropes of science history, the suggestions for the manuscript, and for the wealth of literature suggestions.

David Vandermeulen for the artwork and the open mind toward the relationship between art and science.

Lindy Divarci at MPIWG for the guidance during the publishing process.

Elizabeth Hughes at MPIWG for the critical editing that has made the text clearer and more coherent.

Dr. Peter Gölitz, Editor Emeritus, Angewandte Chemie, for the detailed commentary and scientific insight.

Prof. Gerhard Ertl at FHI, himself a fixture in the history of ammonia synthesis, for the conversations and pithy advice and suggestions.

Prof. Robert Schlögl at FHI and Prof. Jürgen Renn at MPIWG for the years of support and the chance to take on a truly interdisciplinary challenge. This book is a result of your unique relationship.

Most importantly, I thank Prof. Ursula Klein at MPIWG for her experience, expertise, and the patience to turn a physicist into a historian.

Contents

Part I
A Confluence of Factors

Chapter 1
The Object of Investigation

The discovery of ammonia synthesis from the elements forms the basis of discussion for many topics including fertilizer and food, environmental protection, the repercussions of scientific research, and economic transformation, as well as other industrial, political, and social events. The story in this book focuses on the development of the natural sciences and delves into the details of the scientific history of ammonia synthesis. The resulting narrative provides a basis for yet another discussion: the interaction between scientists and how they proceed in their scientific endeavor. The account helps us better contextualize the completion of the scientific research on ammonia and frame the ensuing technological developments along with the consequences for our world, both constructive and destructive.

Today, our lives are overwhelmingly influenced by technological and scientific breakthroughs. The laser, transistor, integrated circuit, and the touch screen have effected change in health care, communications, and even our wait at the supermarket check-out line. While the consequences of such innovations are numerous and easy to identify, the origin of these achievements is obscured or more often simplified as a single event in the history of science. A scientific journal may publish an article, signaling the arrival of a discovery or method, but can the publication of this information be considered the birth of the new knowledge? Can it be reduced to one final result? It seems more prudent and informative—not to mention historically accurate—to describe a discovery in terms of an extended context. Prior achievements and conceptual development gradually lead to a setting in which the discovery is possible. The changes solidify, becoming more profound and particular to the discovery as it nears. Following the breakthrough, further scientific and technological advances appear along with consumer products, all of which become increasingly intertwined with our daily lives; what was previously possible only in the laboratory becomes accessible to the broader public. The connection between the before and after of a scientific breakthrough is more than a single, isolated event; rather, it is a complex and protracted process. This book, in part, focuses on our perception and definition of this connection. We examine it in the framework of

© The Author(s) 2022
B. Johnson, *Making Ammonia*, https://doi.org/10.1007/978-3-030-85532-1_1

the prior, confluent factors as well as the subsequent, expanding consequences—but also as an object of investigation in itself.

An exhaustive historical account would involve a sub-study of every element leading to and extending from a discovery, resulting in an exponential growth of information. This approach is obviously not useful. However, giving a theoretical structure to the vast number of factors helps us selectively delineate the contributing events for clarity. The lines of demarcation may be dynamic and require an awareness of the collective continuity and flow of events in order to be drawn in an informative way (Rudwick 1985, p. 13). Of the many conceivable historical pathways, we can elucidate and contemplate the one which occurred–it is often the result of imperfect or stochastic factors. In some cases, scientific advancement may have been achieved without mature conceptual understanding. In other cases, theories or even entire disciplines may have given way to later advances only to reemerge with new relevance.[1] The dynamic is driven by knowledge moving from one group to another through strategically placed actors or by random interactions. Whatever the specific events, cycles of reapplication result in improvement as the continued interlinking of knowledge leads to discovery.

Here we may play with the question: At what point can new information be considered "known"? Or, alternatively, how much knowledge must be obtained for the term "discovery" to apply (Feynman 1983)? What, despite continuous interlinking, separates a discovery from earlier work contributing to it? Usually, a new discovery can be considered distinct from a prior discovery if the new one is sufficiently more advanced (Kuhn 1970, p. 21). In this case, concepts and physical quantities only hinted at in the earlier context are matured and play a decisive role in the later development. In other cases, older insight may simply become irrelevant or be ignored (Fleck 1980, p. 29). With enough hindsight, we can understand the complexities that were originally difficult to overcome–often they appear less formidable to us today. This advantage provides an optimistic perspective on the challenges we now face, as well as how to identify and realize the technological solutions that lie in our own future. Here, I specifically mention the energy transition (Bruckner et al. 2014; Schlögl 2015a) and the interdependence of the nitrogen and carbon cycles in our biosphere. Though these subjects have been topics of research for decades, they remain issues of concern in need of systemic solutions (Andres et al. 2012; Bolin 1970; Delwiche 1970; Hatfield and Follet 2008; Keeling 1973).

Through consideration of these questions and concerns, we return to ammonia synthesis from the elements, now as a suitable historical object of investigation (Schwarte 1920, pp. 537–551), (Ertl 2012; Scherer 2015; Travis 1993b). At first glance, the process of NH_3 formation from nitrogen (N) and hydrogen (H) appears to be the solution to a simple chemical problem, analogous to the synthesis of water (H_2O). In fact, the process is much more complex. Although ammonia synthesis was already known at the beginning of the nineteenth century, the complexity of the process meant that a large-scale industrial solution was first found over one

---

[1] As an example, see the development of Maxwell's equations in Yeang (2014, chapter 11).

hundred years later. In the first decade of the twentieth century, Fritz Haber, aided experimentally by Gabriel van Oordt and Robert Le Rossignol and theoretically by Walther Nernst,[2] determined the conditions under which ammonia could be directly synthesized from its elements. Several years later, Carl Bosch and Alwin Mittasch were able to upscale the process to an industrial level at the chemical company Badische Anilin- & Soda-Fabrik (BASF). It allowed the mass-manufacture of ammonia to an unprecedented degree.

Ammonia synthesis plays a central role in the most important human endeavor: the production of food. At the beginning of the twentieth century, the agricultural industry's ability to feed the world's population, which had grown significantly during the previous one hundred years, was dependent on natural nitrogen-based fertilizer produced mainly from Chile saltpeter and ammonium sulfate from coking operations. These stock resources were not only limited in quantity, but the reliance on imported sodium nitrate from South America was geo-strategically precarious: ammonia can also be used to manufacture explosives. The country cut off from the natural supply would find itself at an immediate disadvantage. A resource flow was needed.

The advent of synthetic ammonia production meant that a chemical-industrial solution had become available, drastically changing our world. Today, our ability to provide enough nourishment to sustain the world's population is thoroughly dependent on this development. An estimated 30–50% of the current population[3] (and one-third of all those that have lived since 1908) are, put bluntly, able to survive thanks to the insights and developments of modern chemistry. Synthetic fixed nitrogen also impacts the supply of bioenergy (currently about 10% of total global energy). Furthermore, it has enabled a new level of warfare through the manufacture of explosives (Smil 2001, p. *xv*), (Erisman et al. 2008; Stewart et al. 2005), (Sutton et al. 2011c, p. *xxviii*), (Sutton et al. 2011b, p. 33).

Natural growth processes are no longer only geological or biological; they have been replaced by technical-industrial events that are dependent on the results of basic scientific research.

Taking a step back to generalize the context, Haber's work is part of the larger human pursuit of harvesting increasing amounts of energy from our main, and at one point only provider: the sun (Kleidon 2016).[4] His discovery made an important contribution to the transition from draining a stock energy resource to establishing a permanent resource flow. It comprises a transformation of knowledge and a corresponding transformation of societal structures (Renn et al. 2017). Beginning with the control of fire and the neolithic agricultural revolution, humans have endeavored to increase the energy available through photosynthesis, the largest biological mechanism for capturing solar energy on earth (Malanima 2009, pp. 49–

---

[2] Other researchers also played key roles in the development of ammonia synthesis and will be introduced over the course of the book.

[3] This is a global average with local numbers varying by region and diet.

[4] Geothermal and tidal energy provided comparatively negligible contributions.

54). However, supplies remained limited and the ability to harvest crops and perform other work was dependent on human muscle. Draft animals (and slaves) later increased labor output but were not an ideal solution because arable land used to grow food for humans was needed to produce animal fodder. Not until the use of coal began in the early stages of the Industrial Revolution did humans have access to energy reserves independent of the annual inflow from the sun. Despite the abundance of fossil fuels and their ability to form a more stable energy source, crops were still limited by natural sources of fertilizer, whether local or imported. Haber's discovery changed that. Beginning in the eighteenth century, the growth in number and extent of fuel and food-based energy sources had allowed some workers to specialize in other activities and enabled a societal elite to emerge along with a wider array of goods and expectations (Wrigley 2016, pp. 1–30). Resources were available for investment in research and development that ultimately set the stage for Haber's discovery, but the story did not end there. An acceleration of consumption took place after BASF began to produce ammonia on an industrial scale. Soon, fossil fuels were needed for fertilizer production in what was a form of energy transition. Today, we are increasing the amount of the sun's energy we can gather through wind turbines, solar cells, and improved access to other renewable resources. Still, our current energy transition will be a lengthy, complex process dependent on local environments and other conditions, but it is not without precedent. The development of ammonia synthesis, as a historical reflection, illustrates how scientific progress spans generations and adapts to new circumstances. We can learn from these events and apply the lessons to present and future ventures.

The purpose of Part I of this book is to describe the creation of the scientific arena for discovery in which ammonia synthesis was successfully investigated (Rudwick 1985, p. 10, chapter 2). In doing so, I improve on the existing literature in two ways. First, I add to the historical record by incorporating pertinent aspects of physical chemistry, namely Haber and Nernst's implementation and development of the concept of free energy in chemical reactions. This concept was the central physical quantity in their work and it is indispensable in understanding what they achieved and how they achieved it. Previous works on this topic have mainly done one of the following: focused only on a discussion of physical chemistry in a scientific context, embedded incorrect aspects of physical chemistry into a historical context, or neglected a discussion of the subject altogether.[5] Second, I have organized the narrative of events in such a way as to illustrate the confluence of factors leading to an arena for discovery. It is not a chronological account but rather an illumination of the aggregate nature of prior developments critical to Haber's success. Most of the information is based on secondary sources but is synthesized in a novel way to tell a complete story (Sgourev 2015); it is initially broad before converging to show the transformative nature of the events.

The story begins with the emerging industrialization of agriculture and its increasing evaluation on a scientific basis, aided by parallel developments in

---

[5] A recent exception is Deri Sheppard's biography of Robert Le Rossignol Sheppard (2020).

(organic) chemistry. Solving the "ammonia riddle," however, also required an advancement that influenced both industry and the natural sciences: the birth of physical chemistry as a peripheral field, combining chemistry and physics. Only with the maturity of this theory was Fritz Haber's scientific breakthrough possible, and without his results, the resources of the German chemical industry could not have been exploited to upscale his process. Success at the industrial level was also a consequence of advances in fundamental chemistry, such as catalysis, and the establishment of educational facilities. Part I also includes an overview of Haber's work in the laboratory and industrial production at BASF. This transition supplies preparatory context for Part II, a comprehensive case study of Haber's and Nernst's work and their approaches as scientists. The focus in this central section of the book is on the reconstruction of scientific progress based on the decisive publications on ammonia synthesis; it is a portrayal of the aforementioned "connection" between the time before and after a discovery. Such a detailed review of the developments between 1903 and 1908, rooted in the fundamentals of physical chemistry, is given here for the first time and provides a new description of the interaction between Fritz Haber and Walther Nernst that was indispensable for the resolution of the mystery of ammonia synthesis. The rigorous science-historical background also forms a foundation for the discussion of the consequences of technological upheavals. In the conclusions of both Parts I and II, we begin the conversation. Under which conditions and through which pathways are scientific breakthroughs possible? What is the role of the interaction between scientists? How do distinct factors become interlinked? And what are the consequences for our way of life today? These questions are explored more thoroughly in Part III in a theoretical examination of the nature of scientific discovery called *The Haze*.[6] The questions can and should be asked throughout the book—not just in the sections where they are directly addressed. While the goal need not be to answer them completely, keeping the questions in mind will result in a higher level of engagement.

Now let the story begin

---

[6] The three part structure—(1) setting the scene, (2) case study, (3) theoretical inquiry—is similar to the presentation in Martin Rudwick's *The Great Devonian Controversy* (Rudwick 1985).

Chapter 2
The Scientification of Agriculture

It may seem that ammonia synthesis was a scientific triumph for chemistry that also had unexpected consequences for agriculture, much in line with Max Planck's dictum: "Understanding must precede application (Planck 1919)."[1] However, in historical hindsight things often happen just the opposite: in the case of ammonia synthesis, practice and experience preceded theory. Without observations from agriculture and a general understanding of the role of fertilizer—and of nitrogen and the nitrogen cycle in particular—certain essential chemical insight would not have been considered valuable.

From the Middle Ages until the mid-nineteenth century, agriculture in Europe was increasingly strained, broken periodically by plague or war which limited the growing population. The three-field system was the main strategy to meet the increasing need for nourishment, while the development of farmland from forests, along with land to the East and in the Americas, also alleviated pressure. However, crops were also increasingly transported away from their place of origin so that local fields lost their nutrient base and the use of animal excrement as fertilizer became less effective. It led to a vicious spiral. By the turn of the nineteenth century, the situation was dire in many countries. In Great Britain, despite the ongoing "agricultural revolution," experts were aware of the strain on land resources. The economist Adam Smith, for example, stressed the economic consequences of depleted fields while another economist, Thomas Robert Malthus, warned of the difficulties of feeding an ever-growing population. Germany also experienced challenges: during the first two-thirds of the nineteenth century, the country's population grew by 1%, a threatening rate according to contemporary observers.

---

[1] "Dem Anwenden muss das Erkennen vorausgehen."

© The Author(s) 2022
B. Johnson, *Making Ammonia*, https://doi.org/10.1007/978-3-030-85532-1_2

Agricultural crises and famine remained widespread until the Industrial Revolution keeping about three-quarters of a community's labor force devoted to food production. Difficulties with division of labor kept this ratio high. At the beginning of the nineteenth century, many politicians and economists contended that the problems could not be solved with the tools at hand; population growth would eventually overwhelm the maximum productivity of available arable land. Mid-century Europe, particularly Ireland, witnessed widespread famine. Nevertheless, industrialization and continued optimization of agrarian production limited major disasters—but the strategy was not a permanent fix (Gray 1990), (Wrigley 2004, pp. 212–216), (Vanhaute et al. 2007).

How, then, did agriculture modernize into the efficient and calculable system we have today? Like many developments heavily reliant on chemistry, it began in the second half of the eighteenth century during the "chemical revolution"[2] of Antoine de Lavoisier and later John Dalton (Scholz 1987). Their work enabled the first isolation of elements, including carbon, hydrogen, oxygen, and nitrogen. It was then that researchers, most notably the pharmacist and chemist Carl Wilhelm Scheele, were able to identify organic compounds, and understand that these were common to both plants and animals (Cassebaum 1982). Known to be generally different than inorganic compounds, the study of these substances was, not surprisingly, referred to as "plant and animal chemistry." What followed was an increase in identification and even synthesis of organic materials. A finer understanding of these substances and their constitution also strengthened the idea of conservation of matter in chemical reactions. The distinctions between plant chemistry and animal chemistry blurred as animal respiration was shown to be analogous to combustion: animals "burned" carbon for energy and required plants to replenish their carbon supplies. It was later understood that plants, too, exchange gases with the atmosphere (Gräbe 1920, pp. 1–16), (Browne 1977, pp. 170–171), (Holmes 1985, pp. 91–128, 151–198, 291–326). This increasingly scientific view of plants and animals, and their interactions with their environments, developed along with changing cultural factors.

Agricultural methods in the middle of the seventeenth century, especially in central Europe, were based largely on the dissemination of "useful" or "how-to" through what was called *Hausväterliteratur* or *Hausbücher*. Implemented in the previous century, these books were written for aristocratic heads of estates and based on societal conceptions from antiquity. The household was considered the main societal unit and each strove for individual self-sufficiency. At the end of the century, however, parallel to the developments in plant and animal chemistry, a new emphasis on husbandry emerged. Influenced by the Enlightenment, the literature focused on agriculture, industry, and commerce, asserting the claim that agriculture was part of an overall economy that would not only feed the population, but could also create wealth. As free market capitalism came to Europe, discussions of agricultural strategy entered political circles, marking a shift from pre-liberal to liberal thought.

---

[2] Not all historians agree on the magnitude of this "revolution" or if it was even a paradigm shift (Holmes 1989; Klein 2015b).

Although the ideas did not catch on immediately, the approach to agricultural was becoming increasingly based on material and scientific terms (Jones 2017, pp. 6–25), (Gray 1990).

Apart from frequent food shortages, which threatened to rile the populace, two conflicts were integral in establishing the new views. First, the aftermath of the French Revolution legitimatized new liberal thought. The entrenched feuding interests of the past centuries could be openly scorned and governments were able to break the lord-peasant relationship. The capitalist ideas in England of Adam Smith and Arthur Young began to spread across the continent. Second, the Napoleonic Wars (1803–15) showed the vulnerability of a nation's food supplies and land. Desperate politicians were forced to consider possible solutions that they may not have otherwise entertained. It was in this context that Daniel Albrecht Thaer began working for the Prussian State service in 1804 and set up one of the earliest agricultural stations at Möglin in 1806 (Jones 2017, p. 164). Unlike some of his fellow colleagues in the German lands and the rest of Europe, Thaer embraced the modern approach to agriculture and can be counted among "the advocates of 'rational' husbandry [who] were convinced that the time had now come to equip agriculture with a scientific methodology rooted in experimentation (Jones 2017, p. 27)."

Thaer's was not the first agricultural research station (Lavoisier performed experiments at his estate at Fréchines in the 1770s and 80s), but it was one of the earliest to perform work based on scientific knowledge and profited from the first chemical-based breakthroughs in agricultural science. In 1804, Nicolas de Saussure announced his findings that atmospheric CO_2 provided the main source of food for plants. Combined with his belief in the importance of soil rich in nutrients, this report was an early step toward a mineral theory of plant nutrition (de Saussure 1804). It was also around this time that the cultivation of the sugar beet began, which ultimately helped advance the concept of crop maximization and the acceptance of "artificial" fertilizers (Jones 2017, pp. 171–175). The science of "agronomy," based on the ideas of Thaer and de Saussure, was gaining acceptance. A regular publication, *Archive of Agricultural Chemistry for Thinking Farmers*,[3] appeared committed to agricultural chemistry. Lecture series were also popular: at the Royal Institution in London in 1803 (published in 1813) Humphrey Davy stated (Davy 1813, p. 607),

> Agricultural chemistry has for its objects all those changes in the arrangements of matter connected with the growth and nourishment of plants, the comparative values of their produce as food; the constitution of soils; the manner in which lands are enriched by manure, or rendered fertile by the different processes.

However, he continued,

> [while] agricultural chemistry has not yet received a regular systematic form [. . .] it is scarcely possible to enter upon any investigation in agriculture, without finding it connected, more or less, with doctrines or elucidations derived from chemistry.

---

[3] *Archiv der Agrikulturchemie für denkende Landwirthe.*

A long and fruitful development of knowledge centered on improved field research and science-based education had begun (Harwood 2005, pp. 77–109). At the turn of the nineteenth century, the state of agricultural chemistry was different in Great Britain than on the Continent, though neither region had developed an adequate system (Russel 1966, pp. 66–69). Better analysis methods were needed at the research stations before the quantitative results could be achieved which finally shifted attention from crop rotation to soil improvement and fertilizer. Such methods included the use of control plots and work with dung, mathematical models and improved access to methods determining the composition of soils and plants (Jones 2017, pp. 164–170). Especially the latter received a boost from advances in organic chemistry in the first three decades of the nineteenth century with the development of accurate quantitative analysis and the further isolation of organic substances. Primitive superficial analysis techniques, such as combustion and distillation, were improved upon until reliable quantitative results, that is, the amount of hydrogen, carbon, oxygen, nitrogen, etc., in a sample could be confidently obtained (Fig. 2.1). The culmination of these devices, Justus von Liebigs' *Kaliapparat* from 1831, coincided with the birth of modern research stations across Europe (von Liebig 1837), (Ihde 1964, pp. 89–94, 165–179). It was also a period of economic transition to the post-Malthusian Regime. As a higher standard of living changed household activity, more wealth became available for peripheral activities, such as science and technology in agriculture (Galor and Weil 2000; Gray 1990).

One important research station was Jean-Baptiste Boussingault's facility at Bechelbronn in Alsace where he published results beginning in 1836. Boussingault went beyond chemical analysis of plants and fixed soil content by extending the quantitative analysis of de Saussure to analyze how varying quantities of nutrients impacted plant growth. Among these was nitrogen, the exact role of which had long been a mystery. Boussingault demonstrated the importance of the element's assimilation and replenishment in plants, namely that the amount of nitrogen was essentially proportional to plant growth. It was also during these investigations that he noted the soil's improved nitrogen content after the growth of legumes (Browne 1977, pp. 239–251), (Jones 2017, p. 181).

Another leading investigator and supporter of field research was Carl Sprengel (Sprengel 1819). After assisting Thaer for seven years at research stations in Celle and Möglin, he began teaching agricultural chemistry at the university level in 1829. Sprengel was an early adherent to the mineral theory of plant nutrition and aware of the work of de Saussure. In 1831, he moved to Braunschweig to help establish and run a modern research facility, the Ducal Institute for Agriculture and Forestry,[4] but it never came to fruition. Only small field experiments resulted. In 1839, Sprengel moved to Prussia and the *Pommeranian Economic Society*[5] where he established an agricultural academy. It was here that he made his most significant theoretical and

---

[4] Herzogliches Institut für Land- und Forstwirtschaft.

[5] *Pommersche Ökonomische Gesellschaft.*

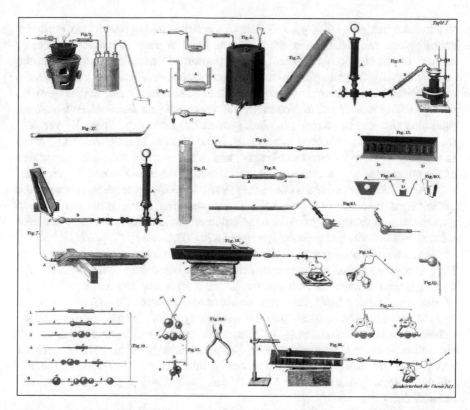

Fig. 2.1 Instruments for determining the composition of substances (von Liebig 1837). Hydrogen is converted to water and carbon to different oxides through combustion (the combination of the organic components of the sample with oxygen). Figures 1–4 show parts of an apparatus for removing water from samples prior to combustion. Figure 5 allows the removal of air from around samples that are especially difficult to dry. The dried sample can then be weighed in the small tube in Fig. 6. The apparatus in Fig. 10 also allows the sample to be heated and dried at reduced pressures. It is already in the combustion tube, C, where it is mixed with copper oxide prior to combustion. Water created during combustion is collected in the calcium chloride tube (Fig. 8) which is attached directly to the combustion tube (Fig. 9). An alternative form of the combined combustion-calcium chloride tube is shown in Fig. 10. Oxidized carbon is captured in the Kaliapparat in Fig. 11; its components, assembly, and preparation are sketched in Figs. 12–14. Figure 15 shows a view from above of the iron sheet oven in which combustion takes place, and Fig. 16 shows the front view. The full assemblage appears in Figs. 18 and 21. *Source*: Niedersächsische Staats- und Universitätsbibliothek, Göttingen, signature: 8 CHEM II 5351

educational contributions to agricultural chemistry (Wendt 1950), (Browne 1977, pp. 231–239), (Frielinghaus and Dalchow 2003).

One of Sprengel's most controversial contentions was the rejection of long-standing beliefs regarding the role of *humus* (decayed organic matter in the top soil) in plant growth. The theory contending the importance of humus can be traced to ancient alchemical ideas and was supported by evidence that manure and plant matter had a positive effect on crops (Brock 1992, pp. 14–20), (Bensaude-

Vincent and Stengers 1996, pp. 1–24). It was believed that humus was the primary contributor to soil fertility and, therefore, the basis of plant nutrition, specifically because of it's high carbon content. Plants purportedly extracted humus from the earth and changed it into plant tissue through combination with water. Supporters of the humus theory thought minerals played, at most, a minor role in plant growth, perhaps acting as a stimulant. The belief persisted well into the nineteenth century, even after the role of minerals in plant growth had been experimentally verified. De Saussure himself believed that humus, in addition to atmospheric CO_2, acted as a source of carbon as well as nitrogen, and other essential nutrients. Supporters of the mineral theory, in contrast, believed plants could be nourished, in principle, with purely inorganic minerals (Waksman 1942). In 1826, Sprengel made one of the earliest clear rejections of the humus theory, arguing that nutrients needed for plant growth are completely supplied from the environment and the lack of any essential mineral will limit further growth (Sprengel 1828), (Browne 1977, p. 231).

De Saussure, Boussingault, and Sprengel did much to advance the mineral theory. They firmly established that the fertility of farmland could only be maintained if the mineral nutrients removed with the harvest were replaced. The actual mechanism of plant nourishment itself, however, remained unexplained. The common attitude was that the nitrogen in plants was drawn directly from diatomic nitrogen in the atmosphere. The mineral theory of plant nourishment was not yet complete. It did not achieve its breakthrough until further knowledge of the role of nitrogen was gained and Justus von Liebig pushed it into the mainstream of agriculture and agricultural education in the 1840s (Finck 2003; Gorham 1991; Gottwald and Schmidt 2003).

Liebig was born in 1803 and received an appointment as professor at the Ludwig's University in Giessen in 1824 at age of 21. He deeply influenced German education in the field of chemistry by establishing the now famous "Liebig School,"[6] which still embodies the turn toward increasing competency in chemistry education in the first half of the nineteenth century (Scholz 1987). The school's effect on the development of chemists and the chemical industry lasted for generations.

Between 1839 and 1845 Liebig and Sprengel published several editions of seminal works in the agrarian sciences. Their regular emphasis on science, especially chemistry, in developing larger crop yields as well as their reliance on experimental observation established definitively the modern mineral theory of agriculture. In 1839, Sprengel published *Die Lehre vom Dünger* (*Fertilizer Studies*, Fig. 2.2, left) with a second expanded edition appearing in 1845 (Sprengel 1839, 1845). Liebig, on the other hand, published the first edition of his *Agrikulturchemie* (*Agricultural Chemistry*) in 1840. However, it was only after the publication of his *Thier-Chemie* (*Animal Chemistry*) (von Liebig 1842) that he felt comfortable enough in his knowledge of dietary cycles of animals and their waste to publish the fifth and definitive edition of *Agrikulturchemie* in 1843 (von Liebig 1843, p. 7) (Fig. 2.2, right).

---

[6] Liebig'sche Schule.

Fig. 2.2 Title pages from (left side) Carl Sprengel's *Die Lehre vom Dünger* (*Fertilizer Studies*, first edition, 1839) (Sprengel 1839) *Source*: Humboldt Universität zu Berlin, signature: AD CA 1986, photographed by the author; and (right side) Justus von Liebig's *Agrikulturchemie* (*Agricultural Chemistry*, fifth edition, 1843) (von Liebig 1843) *Source*: Bayerische Staatsbibliothek München, signature: Chem. 210 he, page 7, urn: nbn:de:bvb:12-bsb10073229-2

At the time, it was known the combustion of plants produced ashes containing inorganic compounds, such as carbon dioxide, ammonia and certain acidic salts, including alkalines combined with silicic acid or phosphoric acid. As Liebig remarked, some (presumably supporters of the humus theory) thought the presence of these inorganic compounds was accidental and they were not critical to a plant's growth (von Liebig 1843, p. 83). However, both Liebig and Sprengel asserted that plants needed these inorganic chemicals to survive and bloom. Furthermore, they took a different form in plants than either before assimilation or in the ashes. In other words, organic plant tissue (and thus animal tissue) was created from purely inorganic compounds and returned to that form after the plant decayed; it explained the cycle of transformation between organic and inorganic matter. Humus could bolster plant growth but was not a necessity. The organic matter in plants did not require its nutrients, or building blocks, to be of similar organic origin. The plants required only a stable, porous medium (soil) through which water could transport dissolved salts. "If we fortify the soil [which already contains silicic acid]," Liebig wrote, "with ammonia and phosphoric acid, which is essential for wheat...we will have fulfilled the conditions necessary for a plentiful harvest because the atmosphere represents an endless reserve of carbon dioxide (von Liebig 1843, p. 174)."[7]

Even the combative and possessive Liebig saw himself as the beneficiary of a European tradition that viewed agriculture not only as a necessity for continued human civilization, but also as a problem with a potentially exact scientific solution. He dedicated his book *Agricultural Chemistry* to his former mentor Alexander von Humboldt (von Liebig 1843, pp. III-IV),

> I don't know whether any part [of *Agricultural Chemistry*] actually belongs to me; when I read the introduction which you gave 42 years ago for Ingenhousz' work "On the Food of Plants," it appears to me, as if I only furthered the opinions which [you] stated and justified therein."[8]

While this may have been customary language at the time, Humboldt indeed foresaw the emergence of the mineral theory. In 1798 he wrote the introduction to the German edition of Jan Ingenhousz' *Essay on the Food of Plants and the Renovation of Soils*. Humboldt's scientific interest at this early stage of agricultural chemistry and plant physiology was not only a matter of improving harvests. Any new understanding of nature, he wrote, "will be beneficial enough if it teaches mankind to decide between competing methods, to clarify everyday yet still incomprehensible phenomena, and to gain insight into the causal connection between outcomes...(Ingenhousz 1798, p. 7)"[9] Humboldt viewed agricultural science not only as a means to achieve technical application, but also as basic research: the information must be collected, regardless of its immediate value. Prophetically, Humboldt touched on how soil composition effects plant growth, and that rain water (which contains nitrogen-based acids) bolsters growth, whereas spring or river water does not. He also stressed the interdependence of animate and inanimate matter.

Liebig reached similar conclusions and was so sure of the consequences he attempted, as did John Bennet Lawes, to market his own line of chemical fertilizer (Cushman 2013, pp. 50–51), (Jones 2017, pp. 181–184). Although the endeavor failed (during the famine of the 1840s) his scientific achievement remained clear. "No one can deny any longer," Liebig wrote, "that further progress in agriculture

---

[7] "Geben wir dem Boden [der schon Kieselsäure enthält] Ammoniak und die den Getreidepflanzen unentbehrlichen phosphorsäueren Salze...so haben wir alle Bedingungen zu einer reichlichen Ernte erfüllt, denn die Atmosphäre ist ein ganz unerschöpfliches Magazin von Kohlensäure."

[8] "...ich weiß nicht ob ein Theil [von *Agriculturchemie*] mir als Eigenthum angehört; wenn ich die Einleitung lese, die Sie vor 42 Jahren zu Ingenhouß Schrift "Über die Ernährung der Pflanzen" gegeben haben, so scheint es mir, als ob ich eigentlich nur die Ansichten weiter ausgeführt hätte, welche [Sie] darin ausgesprochen und begründet [haben]."

[9] "So wird sie [die durch seine Forschung gewonnene Naturkenntnis] doch wohltätig genug für die Menschheit seyn, wenn sie unter entgegen gesetzten Methoden wählen, die alltäglichen, aber noch immer unenträthselten Phänomene erklären, und einen causalen Zusammenhang zwischen Wirkungen einsehen lehrt..."

can only be expected from the field of chemistry now that the conditions for fertile soil and the support of plant life have been determined (von Liebig 1843, p. XII)."[10]

However, many of Liebig's conclusions met with resistance. One notable detractor was Jöns Jacob Berzelius, a supporter of the humus theory. After Liebig sent Berzelius the first edition of *Agrikulturchemie*, the elder scientist responded with comments on December 11, 1840 (Reschke 1978),

> ...I do not completely agree with you that carbolic acid, ammonia, and water are the actual and sole nourishment of plants and that fertilizer in the soil has no other role than to supply these [compounds]. If this were actually the case, it would be possible to nourish the plant to complete maturation of the seed with these given [compounds] [...] If this thesis cannot be proven through experiment, much of the substance of your teaching would fall away [Berzelius then further critiques Liebig's work]. This way of treating science makes for tempting, entertaining reading, but it seems to me an attempt to return to the Fourcroy method, which constructed science from colorful soap bubbles that were swept away after exact examination. From these not even a drop of soap remains from which the bubbles were formed.[11]

However, the influence of Berzelius and other supporters of the humus theory had begun to wane after they made the mistake of attributing a unique chemical nature to each of the many organic compounds found in humus (Kononova 1966). Liebig's reputation, contrarily, was improving, as his work had found many supporters (Jones 2017, p. 184). In 1848, for example, his ideas appeared in the section *Fertilizer Studies*[12] in Karl Nikolaus Fraas' *Historical-encyclopedic Compendium of Agricultural Economics*,[13] though not without some criticism (Fraas 1848, pp. 92–111). While Liebig's *Agrikulturchemie* is often considered the turning point in how scientists, farmers, and politicians viewed agriculture and nitrogen, this attribution is partly a modern one; actual change was not abrupt. The dispute about plant nourishment remained in agricultural science for some time, while confusion (if not outright dissent) abounded. In 1864, 20 years after Liebig's publication, Wilhelm Schumacher, then assistant professor at the Agricultural University of Berlin, wrote (Schumacher 1864, pp. 71–72),

---

[10] "Niemand möchte wohl jetzt, wo die Bedingungen, welche den Boden fruchtbar und fähig machen, das Leben der Pflanzen zu unterhalten, ermittelt sind, leugnen, daß nur von der Chemie aus weitere Fortschritte in der Agricultur erwartet werden können."

[11] "...ich [stimme] nicht mit Deiner Meinung ein, dass Kohlensäure, Ammoniak und Wasser die eigentlichen und ausschliesslichen Nahrungs-Stoffe der Pflanzen sind, und dass der Dünger im Erdboden keine andere Rolle hat als diese hervorzubringen. Wäre das in der That der Fall, so könnte man die Pflanzen mit diesen, schon gebildet, bis zum völligen Reifen der Samen ernähren [...] Wenn nun aber dieser Satz durch Versuche nicht bewiesen werden könnte, so würde viel von Deinem Lehrgebäude wegfallen [Es folgen Kritiken an Liebigs Arbeit]. Diese Art die Wissenschaft zu behandeln, giebt eine hinführende, unterhaltende Lesung, sie scheint mir aber zu der Fourcroy'schen Methode zurückgehen zu wollen, der [sic] die Wissenschaft aus farbenspielenden Seifenblasen aufbaute, welche von der genauen Prüfung weggeblasen wurden, und wovon nicht einmal der Seifentropfen, woraus sie bestanden, zurückgeblieben ist."

[12] *Düngerlehre*.

[13] *Historisch-Enzyklopädischer Grundriß der Landwirthschaftslehre*.

Whoever thinks back a few years will remember the unresolved, yet abandoned dispute between the mineral theorists and the nitrogen theorists [who thought nitrogen was the limiting factor in plant growth]. The "nitrogenists," through their perspective on the nutrition of plants, emphasized mainly ammonia and nitric acid. Everything that contained nitrogen was, as fertilizer, of the greatest importance. Indeed a famous agricultural chemist went so far as to envision the inert nitrogen in the air as accessible to the methods of the chemist so that one day fertilizer could be delivered to farmers through an atmospheric synthetic fertilizer factory. However, this was only an idea borne of inexperience; and yet, the newest science has shown us that such control of the inert atmospheric nitrogen does not belong to the realm of impossibility. Nature itself can, through simple processes, convert nitrogen into ammonia and nitric acid. Luckily, these methods of nitrogen conversion cannot, in general, be made useful for agriculture; I say luckily because the agricultural chemical industry and fertilizer industry could easily overlook the economic interests of preserving atmospheric nitrogen and pilfer the atmosphere until the air is unpalatable for future generations. The current state of the human body, and of the bodies of animals, would be difficult to maintain under a different composition of air with regard to nitrogen and oxygen. And the composition of the atmosphere would certainly change if it were robbed of its nitrogen over thousands of years without otherwise being replenished.[14]

While the "famous agricultural chemist" showed much foresight, Schuhmacher's imagination seemed limited to natural processes of nitrogen fixation (these processes are bacterial and will be discussed shortly). He also overestimated the danger of extracting nitrogen from the atmosphere, which illustrates that the mechanism for replenishing nitrogen in the atmosphere had not been universally accepted or understood–but it did not remain so for long. This mechanism is part of the nitrogen cycle which, along with the law of the minimum, represents Sprengel and Liebig's essential contribution to the understanding of fixed nitrogen in agriculture.

What is now known as *Liebig's law of the minimum* was actually formulated by Sprengel beginning in 1826 and by Liebig in 1840. The law states that when a

---

[14] Wer ein paar Jahre zurückdenkt, wird sich des heute ungelöst beigelegten Streites zwischen Mineral- und Stickstofftheorie erinnern. Die Stickstöffler legten in ihren Anschauungen über die Ernährung der Pflanzen einen Hauptwerth auf das Ammoniak und die Salpetersäure, und Alles hatte als Düngstoff für sie die grösste Wichtigkeit, was nur Stickstoff enthielt, ja ein namhafter Agriculturchemiker ging sogar so weit, dass er den indifferenten Stickstoff der Luft der Kunst des Chemikers zugänglich gemacht wünschte, um dereinst durch eine atmosphärische Kunstdüngerfabrik den stickstoffhungrigen Landwirthen Dünger liefern zu können. Es war dieser Gedanke nur der Traum eines jungen Geistes; dennoch hat die neueste Wissenschaft uns gezeigt, dass eine solche Bemächtigung des indifferenten atmosphärischen Stickstoffes nicht zu den Unmöglichkeiten gehört, dass die Natur selbst durch einfache Prozesse den Stickstoff in Ammoniak und Salpetersäure verwandelt. Glücklicherweise ist die von der Natur gelehrte Methode der Stickstoffüberführung eine solche, dass sie im Grossen für die Landwirthschaft nicht nutzbar gemacht werden kann; glücklicherweise sage ich, weil die technische Agriculturchemie, die Kunstdüngerfabrikation, sehr leicht das Interesse, welches die Volkswirthschaft an der Erhaltung des atmosphärischen Stickstoffes hat, übersehen könnte und durch Beraubung der Atmosphäre deren Luft für künftige Generationen ungeniessbar machen würde: denn die Organisation des menschlichen und überhaupt des thierischen Körpers würde bei einer anderen Zusammensetzung der Luft in Bezug auf Stickstoff und Sauerstoff schwerlich existiren können und die Zusammensetzung der Atmosphäre müsste sich unfehlbar ändern, wenn der Atmosphäre tausende Jahre hindurch Stickstoff entzogen würde, ohne anderweitig Ersatz dafür zu erhalten.

specific element or compound in arable land is limited, the addition of other minerals will have no effect on crop growth on that land. A visualization for the law of the minimum is Liebig's barrel: every plank in the barrel represents an element and the shortest plank limits the water level in the barrel just as that element would limit the growth of a plant in a field. The law of the minimum is now considered the basis for agricultural chemistry and the fundamental idea that agricultural optimization is a quantitative problem. The growth of crops had come to be considered a question of resource inflow in a dynamic which is part of the cycle of elements, especially nitrogen, through the biosphere. In *Agrikulturchemie* Liebig described a cycle of life and death, decay and rebirth by broadening the concept of the nitrogen cycle and emphasizing the *chemical form* of the nitrogen at different points in the process. However, it took several more decades before all of the steps were understood.

Liebig concluded that all nitrogen absorbed by plants came from the atmosphere and was delivered solely via precipitation. Experiments on fertilizer had shown its nitrogen content was limited and it was clear that the vast amount of nitrogen in air was chemically inert. It must be present in another more reactive form when absorbed by plants: nitric acid formed from ammonia. Liebig believed ammonia was the final result of putrefaction and that the compound must, therefore, exist everywhere in the atmosphere, albeit in undetectable quantities. Experimentally, he had not been able to show the existence of ammonia in atmospheric air but thought it existed there as a salt with carbonic acid. The salt's solubility in water meant that precipitation carried ammonia in higher concentrations and continuously washed it out of the atmosphere. Ammonia then decayed into nitric acid, the form of nitrogen essential for plant growth (von Liebig 1843, pp. 50–54, 58–62), (Bradfield 1942). "Ammonia in the form of its salts [nitric acid] is what delivers nitrogen to these plants,"[15] and comprises the main part of the plant protein (von Liebig 1843, p. 62). Today, we refer to these compounds as forms of "fixed nitrogen" in which the triple bond between two atmospheric nitrogen atoms has been broken and the nitrogen is bound to other elements.[16]

Although Liebig stated decaying plant and animal matter as well as animal excrement could be used as fertilizer, he believed these offered no new sources of nitrogen. The ammonia formed during decay was simply returning to the form it had taken in the atmosphere before assimilation. Furthermore, Liebig agreed with the incorrect conclusion of Boussingault that the amount of ammonia from decaying organic matter was irrelevant compared to what was constantly available in the atmosphere (von Liebig 1843, p. 53). This conclusion was based on his observations that peas and beans could thrive without the presence of fertilizer of any kind. He did not believe that they could fix nitrogen on their own (von Liebig 1843, p. 290).

---

[15] "Das Ammoniak in seinen Salzen [Salpetersäure] hat also diesen Pflanzen den Stickstoff geliefert."

[16] Prominent examples of fixed nitrogen are ammonia (NH_3), nitric acid (HNO_3), Chile saltpeter ($NaNO_3$), ammonium sulfate (($NH_4)_2SO_4$), and the ammonium ion (NH_4^+).

For Liebig, the nitrogen cycle comprised the following steps: ammonia, naturally present in the atmosphere in undetectable concentrations, enters the soil via precipitation (atmospheric ammonia is dissolved in higher concentrations in rain water) where it is absorbed by plants in the form of saltpeter. The ammonia later reenters the atmosphere through the decay of organic material or evaporation out of the soil. One question Liebig could not answer, however, was where the ammonia came from in the first place (von Liebig 1843, 279–292).

How, then, did the age-old observation of the benefit of fertilizer on plant growth—a practice at least as old as agriculture itself—fit into the process? Dung and mineral fertilizers had been recycled or brought in from outside ecosystems and used in widespread and varied strategies for thousands of years in Egypt, Peru, and North America. However, logistical difficulties and lack of understanding of its mode of function limited success (Mazoyer and Roudart 2006, pp. 61–64). Although some causal relationships began to emerge around the seventeenth century, conclusions remained dubious. Through trial and error, some farmers were able to optimize their own fields, but their strategies assured nothing for neighboring lands. In some cases, special salt mixtures were known to help. The actual function of fertilizer only began to emerge in the middle of the eighteenth century when it was tied to increased knowledge of the composition of soils. The definitive discovery was made in England where, despite little acceptance of the mineral theory early on, significant scientific strides had been made by mid-century (Jones 2017, p. 170, 182).

Building on results from Boussingault, John Bennet Lawes, an English landowner, and Joseph Henry Gilbert, a chemist from the Liebig School, began large scale experiments at Lawes' agricultural station in Rothamsted, England in 1843 (Fig. 2.3). The investigations ranged from several years to more than a decade and examined the effect of differing amounts of fertilizer on wheat harvests. Through the use of nitrogenous manures containing ammonia salts, they were able to increase the total amount of nitrogen in crop yields by about 40% compared to crops receiving no fertilizer (Lawes et al. 1861). Despite Liebig's authority and staunch defense of his belief, some experimental conclusions contradicted his position and the question of whether plants received all of their nitrogen from the atmosphere remained a matter of debate. Lawes and Gilbert finally proved that plants were able to receive a large portion of their nitrogen from sources that were altogether non-atmospheric. The result uncovered the existence of a new channel into the nitrogen cycle: fertilizer. Crops, indeed, required fixed nitrogen, but plants such as wheat could neither fix the element from the atmosphere themselves, nor did atmospheric ammonia satiate them.

Bolstered by this knowledge, agricultural production in Europe and the United States had managed to stave off the expected famine in the middle of the nineteenth century, but the expansion had made farmers crucially reliant on nitrate imports. The main source of this fixed nitrogen in fertilizer was saltpeter from Chile and guano (millennia-old deposits of sea bird excrement) from Peru (Fig. 2.4) (Honcamp 1930, pp. 3–21), (Leigh 2004, pp. 77–80), (Cushman 2013, chapters 1, 2), (Slotta 2015). Used for centuries in South America, guano first came to

Fig. 2.3 Aerial photograph of Hoosfield at Rothamsted from 1925 showing parceled land. Lawes and Gilbert began an experiment studying the effects of nutrients on spring barley here in 1852. The experiment still looks much the same as it did when it was started. *Source:* Rothamsted Research

Europe in 1803 after Alexander von Humboldt and French botanist Aimé Bonpland returned with samples gathered during a research trip in Peru. However, it was not until two decades later, after global unrest that led to the independence of much of Spanish America, that interest grew and its chemical makeup was assessed: guano contained large amounts of nitrogen. Furthermore, it was soluble in water, making it fast acting. Commercial imports began in the mid-1830s but only became profitable the following decade. Over the next 40 years, the guano reserves were intensively harvested in Peru, mainly for export and often under harsh conditions for the laborers (Clarke and Foster 2009). Although many countries, including France, Germany, Brazil, and Australia, had their own import markets, the chief destinations were Great Britain and the United States. In America, the imports were spurred by competition between farms in the East and new land obtained during westward expansion, but the high prices limited availability. The results of the substance on farm fields in Great Britain, however, were so impressive that guano became their most important fertilizer; imports to Britain climbed to 95,000 tons in 1850 and 300,000 tons in 1858. The international "guano rush" was also helped by outspoken support among prominent farmers and agriculturalists, including in Liebig's magazine *Organic Chemistry*. Worldwide guano imports amounted to 342,000 tons in 1856 and 522,000 tons in 1871 before the best deposits were

NEGATIVE BY M. MOULTON.   Entered according to Act of Congress, in the year 1865, by Alex. Gardner, in the Clerk's Office of the District Court of the District of Columbia.   POSITIVE BY A. GARDNER.

Rays of Sunlight from South America.

STRATA OF GUANO, CHINCHA ISLANDS.

Published by Phile & Solomons, Washington, D. C.

Fig. 2.4 "Strata of Guano, Chincha Islands" from *Rays of Sunlight from South America* by Alexander Gardner (1865). *Source:* The New York Public Library Digital Collections

depleted; shipments declined over the next two decades as the quality of the guano waned and it was eclipsed by nitrates from Chile. The falloff was exacerbated by the mineral theory and its emphasis on the role of nitrogen, more readily available in saltpeter. Farmers also turned to local sources while the Kali Syndikat in Europe and similar organizations in the United States further decreased demand (Mathew 1862), (Skaggs 1994, pp. 1–15, 150–152).

Natural and man-made saltpeter, on the other hand had been traded throughout the world for centuries. It was first used in Asia, before spreading westward via the Dutch East India Company. The European markets consumed much of it for the production of gun powder but its fertilizing power was not realized until the beginning of the nineteenth century (Malanima 2009, p. 64). As demand grew, imports from East India initially dominated the market until imports from South America, especially from the mountains of northern Chile, overwhelmed those from

Asia. The process had begun between 1809 and 1812 when a Bohemian researcher, Tadeáš Haenke, developed a method for extracting saltpeter from what was then Peruvian territory. The reputation of saltpeter as a fertilizer quickly caused the small mining operation to grow and exports began in earnest in the 1830s. The extraction in 1812 was about 1000 tons and grew to 23,000 tons in 1850, and 330,000 tons in 1875 by which time the largest company was Chilean and English-owned. Amidst growing instability in the international nitrate market, tensions between Chile and allied Bolivia and Peru led to the War of the Pacific; nitrate output declined until 1883, when Chile won the saltpeter province of Tarapaca. Chile then held a monopoly on the world nitrate market as exports reached into the millions of tons. The importance of nitrates extended beyond their use as fertilizer: they were also a key ingredient in the production of sulfuric acid, a substance central to the rise of the chemical industry (discussed later). For example, sulfuric acid was necessary to extract ammonia sulfate from coking operations, an important source of fixed nitrogen. It also found use in the Leblanc process and the production of nitric acid and nitroglycerine. Only after the First World War did the synthetic production of ammonia from atmospheric nitrogen displace Chile saltpeter as the world's main source of fixed nitrogen. As if to underscore the changing times, Germany, the world's largest pre-war consumer of Chile saltpeter, was the place of origin of ammonia synthesis (O'Brien 1982, pp. 9–18).

In the second half of the nineteenth century, Chile saltpeter and guano were used in tandem with continuously improving technical and strategic methods of preparing farmland to reinvigorate the fields of Europe. However, Chile saltpeter and guano were stock supplies—a limited natural resource. An independent and permanent source of fixed nitrogen was needed.

Returning to the nitrogen cycle, the mineral nitrogen stocks of South America had vastly increased the cycle's input from fertilizer as the theoretical picture continued to improve. The final input mechanism, a biological one, would soon remind Europeans of a natural phenomenon humans could not replicate. This was the legume. It had been well established that crop rotations should include legumes, because crops planted after the legume harvest showed increased yields. Some scientists, including Boussingault and Lawes, had already observed that legumes could grow well without the addition of fertilizer and that the earth was enriched with nitrogen after their harvest. The investigations continued and grew in complexity into the mid-1880s, but apart from recommendations on how to take advantage of leguminous crops, the mechanism could not be explained. It was attributed to the legumes' possible ability to assimilate nitrogen or to effect a more favorable distribution of the nitrogen remaining in the soil for the following crop.

Advances in plant physiology, chemistry, and, especially, bacteriology were needed before the question could be answered. From the 1850s to 1870s, a debate arose regarding the nature of legume root nodules. Some thought they were a kind of fungus, others that they were of bacterial origin. Finally, in 1886 at the *Meeting of the Society of German Natural Scientists and Doctors*[17] in Berlin, Hermann Hellriegel announced the answer and published the definitive paper two years later (Hellriegel 1886; Hellriegel and Wilfarth 1888). Legumes were able to assimilate more nitrogen for their growth than was directly available from the soil and the secondary source could be inhibited by soil purification: the presence of bacteria working in symbiosis with the roots' nodules fixed atmospheric (or diatomic) nitrogen (Fred et al. 1932, pp. 1–11).[18] Experimental verification of Hellriegel's results soon followed. Apart from providing a further avenue into the nitrogen cycle, the question of whether plants could fix free nitrogen on their own or whether they only assimilated the fixed nitrogen supplied to them could also be answered: it depended on the plant in question. In fact, the action of fixing atmospheric nitrogen for growth, called *diazotrophy*, is carried out mainly by the bacteria of the genera *frankia* and *rhizobium* in symbiotic relationships with plants. Some of the species in these genera can fix nitrogen without symbiosis, instead forming a small group of free-living diazotrophs. There are also phototrophic diazotrophs, the largest group of which are cyanobacteria, formerly called blue-green algae. They have the ability to store chemical energy through oxygenic photosynthesis as well as fix nitrogen for growth in a spatially separated cell (heterocyst) because the production of oxygen generally inhibits nitrogen fixation. Nitrogen fixing bacteria are found in a wide array of environments including large bodies of water as well as moist, arctic, or arid desert soils (Eady 1992, pp. 534–553), (Stal 2015).

Other bacteria, those responsible for the processes of nitrification and denitrification, are vital to understanding the complete nitrogen cycle. In 1873, the German scientist Alexander Müller had noted that nitrates remained stable in sterile solutions, while they disappeared in solutions with untreated water. Four years later, Theophile Schloesing and Achille Müntz refined these observations through their own experiments (Schloesing and Müntz 1877a,b). They passed sewage through a tube containing a mixture of incinerated sand and chalk, and detected ammonia. After twenty days, they detected nitrates and the ammonia content fell to zero. Upon passing chloroform through the pipe the nitrates disappeared and ammonia was again detected. Heating the tube to 100 °C also stopped the transformation of ammonia into nitrates; the process could be renewed and cycled. In the following years it was confirmed that biological processes were responsible for nitrification of the soil. Ammonia is oxidized through a two-step process into a form plants can assimilate: the bacterium *Nitrosomonas* oxidizes ammonia into nitrites, while *Nitrobacter* oxidizes nitrites to nitrates (McKee 1962, pp. 103–112).

Bacteria also supply the mechanism to reduce nitrates back to diatomic nitrogen returning to the atmosphere. This process, known as denitrification, had been observed in experiments throughout the mid-nineteenth century. In the 1880s, Ulysse Gayon, a professor of chemistry at the University of Bordeaux, and Gabriel Dupetit showed that bacteria, again through a multistage process, convert oxidized

---

[17] *Versammlung Deutscher Naturforscher und Ärzte.*

[18] The fixation results initially in the formation of an ammonium ion.

nitrogen in the soil to diatomic nitrogen in the atmosphere. Their experiments also relied on heating and poisoning bacterial cultures to activate and deactivate the denitrification process (McKee 1962, pp. 116–121). In contrast to nitrification, which is carried out by two main bacteria, denitrification is caused by an array of bacteria, the most common of which are *Pseudomonas*, *Bacillus*, and *Alcaligenes*.

With that, the first incarnation of the nitrogen cycle was established (Fig. 2.5).

Today, rather than a stepwise circular pathway, our understanding of the nitrogen cycle in soils is based on the biogeochemical concept of competition between organisms in which physicochemical forces such as diffusion, emission, and erosion supply the distribution mechanisms. The existence of the cycle itself is based on nitrogen's ability to assume an unusually large number of oxidation states ranging from -3 to +5. The nitrogen in N_2 (oxidation state 0) can gain up to 3 electrons in reduction (fixation) processes or it can lose up to 5 electrons in oxidation (nitrification) processes. While the reduction processes (via enzyme) generally release more energy than the oxidation processes, organisms driving the cycle have evolved to extract energy from each step. The plants themselves can assimilate nitrogen in either fixed form (ammonium ions, NH_4^+) or as nitrates (NO_3^-). In cases where these inorganic compounds are limited, plants may assimilate organic nitrogen in the form of amino groups ($-NH_2$). In other words, local conditions dictate which forms of nitrogen are introduced into the local nitrogen cycle and determine the balance of conversion reactions.

Thus, in contrast to soils where the nitrogen cycle is dominated by bacterial processes, the cycle in the atmosphere is governed mainly by chemical reactions. Here, the input mechanisms are reduced nitrogen in the form of ammonia gas from agriculture and oxidized nitrogen (NO_x) from the burning of fossil fuels. Aquatic, marine, and coastal ecosystems, which have vast contact areas with soil and atmosphere, will have a mixture of characteristics. Knowledge of the intricate pathways in these systems has been increasing since the 1950s, thanks primarily to the use of the ^{15}N tracer isotope (Butterbach-Bahl et al. 2011; Delwiche 1970; Durand et al. 2011; Hertel et al. 2011; Müller and Rehder 2015; Rennenberg et al. 2009; Seitzinger et al. 2015; van Groenigen et al. 2015; Voss et al. 2011).

In Liebig's time, however, the greatest achievement of the study of the chemical basis of agriculture was to show the importance of nitrogen for plants and animals. It was this early grasp of the nitrogen cycle that allowed an understanding of the depth and dynamics of the element's role in the biosphere. As we have seen, the knowledge was accumulated by many researchers over decades of interwoven and overlapping discoveries.

The importance of ammonia's role in food production had far-reaching economic and technological consequences, and also impacted chemistry as a scientific field. This influence was caused by several factors. By the 1880s, the finite sources of fixed nitrogen in South America had been identified as a bottleneck for further socioeconomic development. Replacing these resources became a political, economic, scientific, and technological challenge. Also, the realities of the nitrogen cycle posed fundamental questions about nitrogen fixation, specifically about ammonia synthesis: which biological mechanism did nature perform that chemistry could not

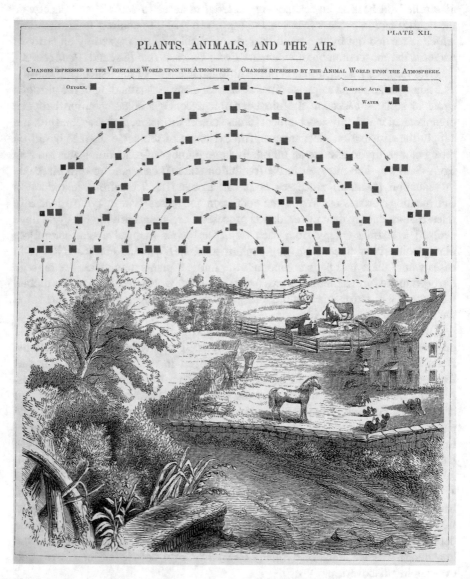

Fig. 2.5 This page: biospheric cycles of oxygen, carbonic acid, and water from *Youmans' Atlas of Chemistry* (1856), illustrating the interdependence of plants and animals (Youmans 1856, p. 87). Fixed nitrogen is not included in the graphic, although the role of ammonia and "neutral nitrogenized compounds" are discussed in the text. Opposite page: the nitrogen cycle according to Delwiche in 1970 (Delwiche 1970). This depiction is still current, and includes the additional step of the nitrogen cascade (Sutton et al. 2011c). Today, the complexity of the nitrogen, carbon, oxygen, water, etc. cycles is understood to such a degree that simplicity can only be preserved if they are shown separately

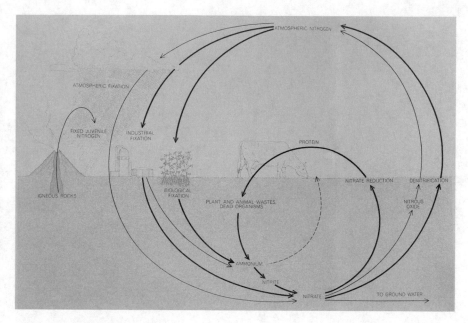

Fig. 2.5 (continued)

mimic? Ammonia synthesis became both an economic and a scientific problem, an interconnection that would continuously encourage development.

To better understand the circumstances of the scientific arena that made ammonia synthesis possible, we now turn to the transformation of organic chemistry from its early empirical stage to its hybridization with physics as well as the comprehension of the role of a catalyst in chemical reactions.

Chapter 3
Advances in Organic Chemistry, Catalysis, and the Chemical Industry

As we saw in the last section, the modern understanding of chemistry began in the second half of the eighteenth century with the advances of Antoine Lavoisier and others, and further contributions from John Dalton and Jöns Jacob Berzelius in the early 1800s. This period saw the emergence of organic chemistry, which, with the help of new conceptual and experimental tools, established itself as a discipline separate from inorganic chemistry over the next half century (Scholz 1987). The researchers of this period encountered considerable confusion due to the complexities and behavior of organic structures. Aided by increasingly accurate elementary analysis, they nevertheless made significant advances based on the empirical studies of structure chemistry. While these developments, along with other factors, helped the chemical and dye industries make large strides in chemical synthesis, the conceptual and experimental tools did not supply sufficient understanding of chemical systems to enable ammonia synthesis from the elements. Here we will focus on two developments that illustrate the experimentally-based approach to organic chemistry in the 1800s in order to frame the conceptual leap in physical chemistry toward the end of the century that led to Fritz Haber's breakthrough.

The first was the formulation by Eilhard Mitscherlich in 1834, after studying the production of ether from ethanol and sulfuric acid (Fig. 3.1), that catalysis represented a specific kind of chemical phenomenon (Mitscherlich 1834b). With this assertion, many scattered observations were brought together under one classification.

The reaction producing ether was well known long before Mitscherlich gave it theoretical underpinning. The first explanation came in 1797 and while different theories emerged over the next 35 years, they often ascribed a chemically active role to the sulfuric acid or suffered other drawbacks. Mitscherlich's experimental setup was not new; he dripped ethanol into a mixture of boiling dilute sulfuric acid and alcohol to produce ether. It was, rather, his novel interpretation of the results which

© The Author(s) 2022
B. Johnson, *Making Ammonia*, https://doi.org/10.1007/978-3-030-85532-1_3

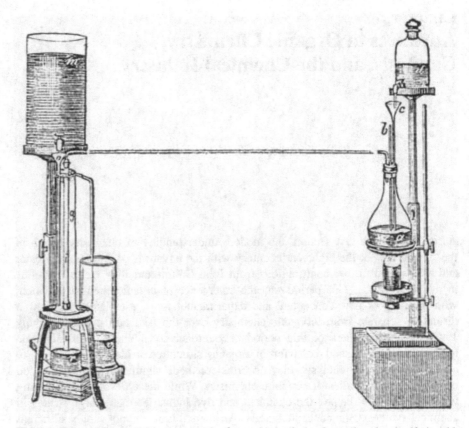

Fig. 3.1 Eilhard Mitscherlich's apparatus for the synthesis of ether from ethanol and sulfuric acid. The ether was produced in the heated flask (a) on the right which contained sulfuric acid, water, and a thermometer (b). The alcohol was contained in the flask (d) and could be continuously supplied. The long pipe joining the left section to the right section of the apparatus served to collect the volatile ether. After the ether was cooled with water from the tank (m) and deposited in the long tube below, it could be removed (Mitscherlich 1834b)

represented the advancement (Gräbe 1920, pp. 94–96), (Mittasch 1939, pp. 30–36), (Schütt 1992, pp. 121–127). "Therefore," wrote Mitscherlich in consideration of his results,

> it follows from the stated facts that alcohol dissociates into ether and water in contact with sulfuric acid at a temperature of about 140°. Decomposition and formation occur very often in this way; we will call them decomposition and formation through contact. The best example is oxidized water; the smallest trace of fmanganese superoxide, of gold, of silver, or of other substances causes a dissociation into water and oxygen gas...without themselves sustaining the slightest alteration. The dissociation of [different] types of sugar in alcohol and carbonic acid, the oxidation of alcohol when it is transformed into acetic acid and the dissociation of urea and water into carbonic acid and ammonia also belong to this type of reaction. Initially, the substances are not changed. However, through the addition of a very slight quantity of ferment, which in this case is the contact substance, at a specific

temperature, the change takes place rapidly. The transformation of starch into starch sugar, when it is boiled in water with sulfuric acid is very similar to the production of ether...[1]

Mitscherlich, for the first time, collected many seemingly different chemical reactions, all of which were promoted by the presence of a substance that remained unchanged during the reaction, under a single expression: decomposition or formation *through contact*.

The importance of Mitscherlich's idea was seized upon by Berzelius who wrote the following year (Berzelius 1835),[2]

> It is then shown that several simple and compound bodies, soluble and insoluble, have the property of exercising on other bodies an action very different from chemical affinity. The body effecting the changes does not take part in the reaction and remains unaltered through the reaction. This unknown body acts by means of an internal force, whose nature is unknown to us. This new force, up till now unknown, is common to organic and inorganic nature. I do not believe that this force is independent of the electrochemical affinities of matter; I believe on the contrary, that it is a new manifestation of the same, but, since we cannot see their connection and independence, it will be more convenient to designate the force by a new name. I will therefore call it the "Catalytic Force" and I will call "Catalysis" the decomposition of bodies by this force, in the same way that we call by "Analysis" the decomposition of bodies by chemical affinity.

Berzelius also recognized that catalysis was an important factor in living organisms and introduced contact processes into plant and animal chemistry.

Neither Berzelius nor Mitscherlich were the first to recognize the inimitability of catalytic processes (Mittasch 1939, pp. 1–36). Catalytic investigations in the era after Lavoisier began in 1781 with the transformation of potato starch into dextrin (a form of sugar) after the addition of tartaric acid or acetic acid. In 1811, Johann Wolfgang Döbereiner noticed the speed of the reaction depended on the concentration of the acid and postulated the reaction would eventually take place even if no acid were present. The boiling water changed the starch into dextrin and the acid only expedited the conversion. Soon after, it was noted that the facilitating substance was itself not changed during the reaction, but it remained unclear whether this behavior was altogether a new type of chemical phenomenon

---

[1] "Aus den angeführten Thatsachen folgt also, daß Alkohol in Berührung mit Schwefelsäure bei einer Temperatur von ungefähr 140° in Aether und Wasser zerfalle. Zersetzung und Verbindung, welche auf diese Weise hervorgebracht werden, kommen sehr häufig vor; wir wollen sie Zersetzung und Verbindung durch Contact nennen. Das schönste Beispiel bietet das oxydirte Wasser dar; die geringste Spur von Mangansuperoxyd, von Gold, von Silber und anderen Substanzen bringt ein Zerfallen der Verbindung in Wasser und Sauerstoffgas...hervor, ohne daß diese Körper die mindeste Veränderung erleiden. Das Zerfallen der Zukerarten in Alkohol und Kohlensäure, die Oxydation des Alkohols, wenn er in Essigsäure umgeändert wird, das Zerfallen des Harnstoffs und des Wassers in Kohlensäure und Ammoniak gehören hierher. Für sich erleiden diese Substanzen keine Veränderung, aber durch den Zusatz einer sehr geringen Menge Ferment, welches dabei die Contactsubstanz ist, und bei einer bestimmten Temperatur, findet diese sogleich statt. Die Umänderung der Stärke in Stärkezucker, wenn man Stärke mit Wasser und Schwefelsäure kocht, ist der Aetherbildung ganz ähnlich..."

[2] Translation from Swedish in Lindström and Pettersson (2003).

and there was little discussion. These and similar experiments investigated what is now called *homogeneous catalysis*, where the reacting substances and the catalyst are in the same material phase. Gaining detailed insight was difficult. In contrast, when observing substances in different phases, now called *heterogeneous catalysis*, the separate components of the reaction were more easily recognized. For example, in 1783, Jospeh Priestly allowed alcohol vapor to flow through a heated tobacco pipe and obtained ethylene.

Experiments on ammonia are also encountered in early catalysis research. One of the first was in 1788 when William Austin found tiny amounts of ammonia ("volatile alkali") when nitrogen ("phlogisticated air") was combined in a glass tube with water and iron filings. The iron in contact with water evolved hydrogen ("light inflammable air"), which, if it met the nitrogen "at the instant of its extrication," combined to form ammonia (Austin 1788), (Thomson 1802, pp. 313–314), (Leigh 2004, p. 104). At the turn of the century, Humphrey Davy dissociated ammonia in a heated glass tube using a copper wire. His experiments were extended by Louis Jacques Thénard in 1813 using a variety of metals: Fe, Cu, Ag, Au, and Pt. Thénard noticed by testing an increased number of metals that there was a wide range of catalytic ability and attributed it to the large heat capacity and heat transfer through some metals. In the case of ammonia, Fe was the most effective catalyst but only the dissociation reaction seemed to proceed spontaneously. Ammonia was found ubiquitously in nature, so why was it, in contrast to water, so difficult to generate catalytically? The similarity in chemical formulae, NH_3 and H_2O, provided no clarification: the mystery of ammonia synthesis had begun.

Also of note in connection with the later success of ammonia synthesis and its role in fertilizer is Davy's realization in 1812 that the presence of NO/NO_2 in the lead chamber process increased the likelihood that sulfur dioxide (SO_2) would be oxidized to sulfuric acid (H_2SO_4) with nitrosylsulfuric acid ($NOHSO_4$) as an intermediate step. It was reasoned that the NO/NO_2 took the place of platinum in heterogeneous reactions and led to the same dissociation and generation through contact. Ammonia could then be reacted with H_2SO_4 to make ammonium sulfate (($NH_4)_2SO_4$), a form of fixed nitrogen plants can assimilate (fertilizer).

In 1816, Davy performed important experiments on platinum, a metal which soon came to be considered the ultimate catalyst. He burned methane and other gases in air in the presence of a platinum wire below its annealing temperature (the wire was heated in the process and strengthened the theory of the heat capacity). The use of potash as an impurity and a finer, extended distribution of platinum were found to be advantageous, but no explanation could be found. Around this time, Döbereiner described catalytic activity as similar to ice crystal seeding, while Thénard suspected there was an electrical component to the mysterious phenomenon. He ventured the statement that all of the new observations may be due to the same "force" and began to consolidate them under a single description. In the 1820s, Döbereiner continued experiments with platinum and platinum sponges as contact materials. When the metal was in the presence of hydrogen, the gas began to combine with oxygen to form water; if oxygen concentrations were too low, the hydrogen combined with nitrogen to form small amounts of ammonia. The result was again puzzling because the large quantities of nitrogen in air did not lead to an appreciable ammonia yield–nitrogen was obstinate in its refusal to react with hydrogen. Despite the unanswered questions, Döbereiner's experiments on platinum contributed greatly to the understanding of catalysis and unleashed a wave of research lasting into the 1840s that charted the temperature dependence of an increasing number of metals.

During his experiments, Döbereiner, too, began to consider many of the new observations to be consequences of the same behavior. The reactions were influenced "via mere contact and without participation by outside powers."[3] However, he and other scientists did not understand the link between dissociation and association; they are competing and dynamic components of any chemical reaction. Not surprisingly, differing and widespread explanations of catalytic phenomena abounded in the 1820s, although it was clear that they were not "classical" chemical reactions. Rather than the results of catalytic reactions, the *nature* of catalytic action was becoming the focus of research, as exemplified in Mitscherlich's work. Despite his and Berzelius' unifying nomenclature, the terms *catalysis* and *contact* were not introduced without friction. Mitscherlich, Berzelius, Liebig, and others had (strongly) differing opinions on catalytic action, and for years the nature of the mechanism underlying the process was contested. Their efforts were, however, strongly influenced by empirical organic chemistry and did not lead to an understanding of the underlying role of a catalyst. With regard to ammonia synthesis in particular, Alwin Mittasch wrote (Mittasch 1951, p. 61),[4]

> Although the number of ammonia catalysts and the conceivable synthesis methods that were investigated were continuously expanded by the efforts of the most diverse investigators during the course of a century, the results would inevitably remain negative as long as the tools of the newer chemical research [physical chemistry] could not be applied to their fullest extent. Empirical research and technology could only reach the ambitious goal [of ammonia synthesis] after the establishment of a strict scientific definition of what was theoretically achievable in this henceforth highly focused field.[5]

While it suffices here to state that catalytic or contact processes were defined as their own class of chemical phenomena in the nineteenth century and that the development of physical chemistry allowed for a deeper understanding of the role they play in chemical reactions (raising both forward and backward reaction rates), the mechanism behind catalytic activity has never shed its air of mystery. Somehow, the contact material *creates* a physical relationship between substances that is

---

[3] "...durch bloße Berührung und ohne alle Mitwirkung äußerer Potenzen."

[4] For a more complete list and discussion of catalysis, especially as applies to ammonia, see Mittasch (1939) and Mittasch (1951).

[5] Obgleich der Kreis der untersuchten Ammoniakkatalysatoren und der denkbaren Arbeitsweisen durch die Anstrengungen der verschiedensten Forscher im Laufe einer hundertjährigen Entwicklung immer mehr erweitert wurde, so mußte das Resultat doch immer wieder ein negatives bleiben, so lange nicht die Hilfsmittel der neueren chemischen Forschung [der physikalischen Chemie] in vollem Umfang eingesetzt werden konnten. Erst nach streng wissenschaftlicher Festlegung des theoretisch überhaupt Erreichbaren konnten auf dem nunmehr stark eingeengten Gebiet die empirische Forschung und schließlich die Technik dazu gelangen, das gesteckte Ziel zu erreichen.

otherwise absent. The individual steps of the ammonia synthesis and other catalytic reactions are well-understood on a physicochemical basis for model catalysts such as iron single crystals. However, there is a material gap between model systems and the performance systems used in industry. No consensus exists on the exact nature of the active centers on performance catalyst surfaces and for this reason, complete catalyst engineering is not yet possible (Campbell 1994; Ertl 1980, 1990, 2008; Ertl et al. 1978, 1981, 1983; Schlögl 2003, 2015b, 2020; Somorjai and Park 2009; Spencer et al. 1981, 1982; Stoltze and Nørskov 1985, 1988). Catalysis research remains an evolving field.

Again using ammonia as an example, it was thought for most of the twentieth century that the catalytic ammonia synthesis reaction was completely dependent on the behavior of N_2. Specifically, its dissociation (breaking the triple bond between the two N atoms) and its ability to adsorb onto the catalyst surface were seen as hurdles. Adsorbed single nitrogen atoms blocked active sites and impeded further N_2 molecules from dissociation; they were the limiting factor for the reaction (Emmett and Brunauer 1933; Ertl et al. 1981). In the past two decades, however, it has emerged that the reaction is much more complex and dependent on microkinetics at the catalyst surface. In addition to the potential energy barrier for N_2 dissociation, a different potential barrier can inhibit the hydrogenation of N to NH, NH_2, and finally NH_3; in general the process grows more difficult over the six hydrogenation steps of two nitrogen atoms and the generation of two molecules of ammonia. Under certain conditions, this barrier may be more limiting than the dissociation of N_2. Furthermore, the ability of NH, NH_2, and NH_3 to adhere to the catalyst surface is of the same order of magnitude as the N atoms and contributes to the satiation of active sites in what is termed *autopoisoning*. These two important characteristics of a catalyst, the ability to dissociate N_2 and to release N, NH, NH_2, and NH_3 from the surface to free active sites, are linked in a specific and, in conventional catalysts, non-ideal way (scaling relations) that limits the efficacy of today's materials (Abild-Pedersen et al. 2007; Dahl et al. 2001; Geng et al. 2018; Honkala et al. 2007; Jacobsen et al. 2001; McKay et al. 2015; Poobalasuntharam et al. 2011; Vojvodic and Nørskov 2015; Vojvodic et al. 2014; Wang et al. 2011).

An overview and modern discussion of catalytic phenomena can be found in Schlögl (2015b).

Returning to organic chemistry, the second development we will discuss is the initial isolation (1825) and production (1834) of benzene—a hydrocarbon (its aromatic (ring) structure was proposed later, in 1865, by August Kekulé). The experimental investigation of benzene provides an example of the empirical approach in the nineteenth century and is also the basis for the aniline molecule in which one hydrogen atom from the benzene ring is replaced by an amino group (-NH_2). Aniline was the first synthetic dyestuff and played a central role in the rise of the German dyestuff industry (Travis 1993a, pp. 166–178).

Dalton had already recognized the individual hydrocarbon compounds *carburetted hydrogen* (methane, CH_4) and *olefiant gas* (ethylene, C_2H_4) at the beginning of the nineteenth century. In 1825, Michael Faraday isolated benzene, which he called *bi-carburetted hydrogen* (Faraday 1825), (Rocke 1984, pp. 29–40), (Ihde 1964, pp. 101–109). However, lack of knowledge about the structure of hydrocarbons prevented an exact stoichiometric identification. It was Mitscherlich who, in 1834, first synthesized benzene from benzoic acid heated over calcium hydroxide and proposed that it was composed of three proportions of carbon and three proportions of hydrogen (Mitscherlich 1834a), (Gräbe 1920, pp. 59–65), (Ihde 1964, pp. 184–190), (Rocke 1984, pp. 174–175), (Schütt 1992, pp. 114–120). Describing his method of examination, Mitscherich wrote,

> If the benzoic acid is brought together with a strong base...and the mixture submitted to distillation, initially water will separate before a thin, oily fluid, develops which floats on top of the water. If the mixture is very slowly warmed, the residue in the retort is completely colorless and leaves no trace if dissolved with acid, [a reaction which] evolves carbonic acid; the solution in the acid is colorless and no gas is evolved during distillation. The benzoic acid disaggregates, therefore, into carbonic acid and an oily liquid [which turned out to be benzene] [...] Because this liquid can be obtained from benzoic acid and is likely related to the benzoyl compound, it should preferably be named benzol [benzene]...[6]

Distillation and combustion were two important methods of elementary analysis used in the nineteenth century to asses the makeup of organic matter (for example, of leaves) (Ihde 1964, pp. 173–179). For contrast, these empirical tools may be compared to the approaches of Haber and Nernst between 1903 and 1908 in the context of physicochemical experiments. Combined with catalysis, the conceptual tools of the empirical period of organic analysis made important contributions to the development of the chemical industry–even if they could not be used to determine a method for ammonia synthesis from the elements. Though it was not the first interaction between chemistry and industry (Scholz 1987), the dyes derived from aniline and other aromatic molecules after about 1860 were crucial in allowing the chemical industry to amass the capital and infrastructure that later allowed for a generous investment in ammonia synthesis research.

Aniline was identified in 1826 and had been isolated several times before August Wilhelm von Hofmann, a student of the Liebig School, clarified its constitution (C_6H_7N) in 1843 (Hofmann 1843; Perkin 1862; Zinin 1842), (Gräbe 1920, pp. 139–147, 213–219). It was during his research into the makeup of coal tar, which was considered a useless and even bothersome byproduct of the coking industry, that he improved the distillation process used to obtain aniline and realized it was available in larger quantities than originally thought. The discovery was important: the economic potential of aniline had already been identified for the production of synthetic (or coal tar) dyes. If aniline could be obtained cheaply it would facilitate

---

[6] Mengt man Benzoësäure mit einer starken Base...und unterwirft das Gemenge der Destillation, so geht zuerst Wasser und zuletzt eine dünnflüssige ölartige Flüssigkeit über, welche auf dem Wasser schwimmt. Wenn man das Gemenge sehr langsam erwärmt, so ist der Rückstand in der Retorte vollkommen farblos, und läßt beim Auflösen in Säure, wobei sich Kohlensäure entwickelt, keine Spur eines Rückstandes zurück; die Auflösung in der Säure ist farblos, und es wird keine Gasart bei der Distillation entwickelt. Die Benzoësäure zerlegt sich also in Kohlensäure und in die ölartige Flüßigkeit [was sich als Benzol herausstellte] [...] Da diese Flüßigkeit aus der Benzoësäure gewonnen wird, und wahrscheinlich mit den Benzoylverbindungen im Zusammenhang steht, so gibt man ihr am besten den Namen Benzol...

the manufacture of these materials: they could replace the often costly biological sources of colorant and transform the growing international textile industry (Travis 1993a, pp. 31–33).

The first aniline-based dyestuff was aniline purple (also called tyrian purple or mauve), produced in 1856 by William Henry Perkin, a student of Hofmann's. At the time, the British Empire was fighting the Crimean War and needed quinine to combat malaria. During his attempts to produce quinine, Perkin employed a simpler reaction involving aniline sulfate and potassium bichromate, which resulted in "a very unpromising black precipitate." The substance, however, contained aniline purple and Perkin was able to isolate it (Perkin 1862). Helped by the Béchamp process for the large scale reduction of nitrobenzene to aniline (Béchamp 1854), it became the first commercially produced dyestuff made from coal tar and essentially started the modern organic chemical industry. In the same year, Perkin patented the production process. Later, he noted the role of pure science in his discovery, a quotation one could easily apply to ammonia synthesis (Perkin 1862).

...the process which is now employed for [the preparation of aniline] is a remarkable instance of the manner in which abstract scientific research becomes in the course of time of the most important practical service.

Perkin's undertaking resulted not only in the development of new technology for chemical treatment and textile printing, but his success led others, including August Wilhelm von Hofmann himself, to enter the coal tar dye industry and draw on Perkin's patented process. While the new dyes were not as durable as the original, they were more brilliant and more popular with the public. Soon, the competition, especially the aniline dye fuchsine, drove mauve from the market, and the new products secured the future of the industry (Edelstein 1961, pp. 759–765), (Travis 1993a, pp. 31–64).

The emergence of the coal tar dye industry brought together men of different backgrounds. Expertise in chemistry, catalysis, the printing trade, and engineering as well as access to capital all shared in making the new industry lucrative. The company BASF, which will appear again during Fritz Haber's work on ammonia, illustrates the successful synthesis of these fields as well as the growth and hybridization of chemistry and industry (Abelshauser et al. 2004, pp. 1–27).

A forerunner company to BASF was founded in Mannheim in 1861 by Friedrich Engelhorn, the son of a brewmaster. Born in Mannheim in 1821, he left his hometown for nine years as a young man to travel Europe and further his education in the jewelry trade. In 1846, he returned as the small city of Mannheim began to feel the effects of the industrial revolution. Engelhorn's experiences during his travels had brought him to European centers transformed by industrialization and, once back in Mannheim, he decided to pursue a career in industry rather than jewelry. In 1848, he founded a company that produced bottled gas for lighting–a mark of modern, industrialized cities. By 1851, his business had expanded and taken over the lease of the Mannheim gas works. Engelhorn himself directed the company both commercially and technically. In 1865, he sold his share in the company to concentrate on the aniline and dyestuff factory he had built next to the gas works several years earlier. Engelhorn had immediate access to a supply of coal tar from the neighboring gas production facility and benefited from the growing chemical infrastructure in Mannheim from which he obtained important precursor chemicals. Engelhorn chose Carl Clemm as technical director of the new company, a chemist who had experience with aniline dyes. Clemm had been recommended by his uncle Carl Clemm-Lennig, the director of the Mannheim Fertilizer Factory and former student of the Liebig School.

While other chemical companies began producing coal tar dyes in addition to their usual commercial products, Engelhorn focused on large-scale dye production from the beginning. Throughout the first half of the 1860s, his company continued to expand its infrastructure and hired Carl Clemm's younger brother August as second technical director. The growth was so swift that Engelhorn soon wondered whether it would not be more profitable to produce the needed inorganic chemical precursors himself rather than continuing to purchase them from Mannheim's established chemical provider, the Association of Chemical Factories.[7] Engelhorn approached the association in 1864 in an attempt to negotiate a better deal for his company, but when the agreement failed he decided to enter into direct competition and manufacture his own chemicals. At this time, the venture was risky because the establishment of other dye companies had made the future profitability of Engelhorn's company, now called Sonntag, Engelhorn & Clemm, anything but certain. The breadth of the undertaking was too cost intensive for the company to spearhead on its own so Engelhorn turned to the Mannheim bank W.H. Ladenburg & Sons for assistance. Engelhorn's standing in the industrial community of Mannheim, along with the reputations of the Clemms and fellow founder Friedrich August Sonntag, helped the partners secure the necessary funds from the bank and other investors to establish the first largely self-sufficient coal tar dye company. In 1865 the Badische Anilin- & Soda-Fabrik (BASF) was born.

Engelhorn had originally wished to keep the company in Mannheim but difficulties in negotiating the purchase of a suitable building site caused the company to look elsewhere. Eventually, he found a site across the Rhine River in Ludswigshafen. The smaller town was excited by the chance to catch up to its larger neighbor on the eastern bank and quickly authorized the purchase of the necessary land. The official seat of BASF remained, however, in Mannheim until 1919. Almost immediately, plans were finalized for new buildings and modern lighting. Rail links were proposed to connect the site to western destinations and the construction of a new bridge over the river connected rail lines to the east. The location also offered water, waste removal, and shipping opportunities from the Rhine. The Ludwigshafen site was ideal (Fig. 3.2).

Engelhorn lost no time initiating his plans of comprehensive dyestuff production on an unprecedented level. The company had been structured to allow access to new capital in the event of expansion, which, combined with the wealth of chemical and commercial experience, the internal production of precursor chemicals, and

---

[7] Verein Chemischer Fabriken

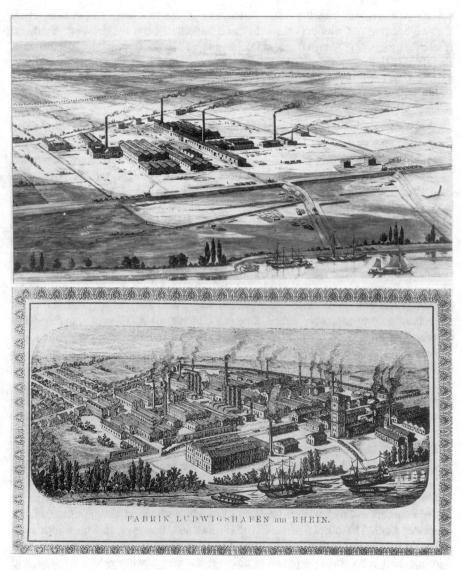

Fig. 3.2 The development of the BASF site at Ludwigshafen (1866–1901). This page, top: a photograph of a painting by Otto Bollhagen completed circa 1922, showing the site as it was circa 1866; bottom: 1873, drawing by W. Menges; 1881; Opposite page, top: painting by Robert Stieler; 1901; bottom: lithograph by Christoph Seitz. Source: BASF SE, Corporate History, Ludwigshafen

geographical advantages, launched BASF into the position of industry leader for decades. During this time, in 1865, it entered into agreements with partner companies and divided the dyestuff market to limit competition and increase efficiency—a practice that became common in the industry. BASF produced only red aniline dyes (fuchsine), which allowed it to invest resources in its inorganic

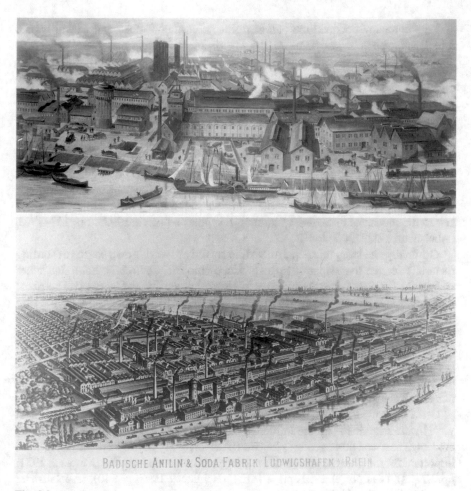

BADISCHE ANILIN & SODA FABRIK LUDWIGSHAFEN RHEIN

Fig. 3.2 (continued)

chemistry department and expand the related infrastructure and expertise. At the end of the 1860s, a theoretical basis, in particular structural formulae, began to strengthen the empirical approach to organic molecule synthesis. Kekulé's 1865 description of benzene as a ring of six carbon atoms was used to describe more complicated molecules and aided in the production of new synthetic dyestuffs. This is exemplified by alizarin, a synthetic replacement of the dye produced from the madder root. In 1868, in Adolf Baeyer's laboratories at the Gewerbe Institute in Berlin, Carl Graebe and Carl Liebermann successfully synthesized alizarin from anthracene and anthraquinone, aided by the new theoretical considerations: all three molecules are fusions of three aromatic carbon rings (with appropriate substitutions of the hydrogen atoms). It was an important step in the synthesis of naturally occurring materials. Alizarin would be important for BASF's continued success and the company obtained the patent rights soon after Graebe and Liebermann's initial achievement, partly due to the efforts of Heinrich Caro. Caro had come to BASF in 1868 and, beyond his connections to W.H. Perkin and others in the English and German chemical industries, had developed organic synthesis methods with a science-based approach that proved valuable to his employer. He was joined in 1869 by Carl Glaser as part of a Friedrich Engelhorn's continued effort to secure scientifically skilled employees. Graebe, Liebermann, Caro and Glaser later improved and upscaled the production of alizarin; the process remained central to BASF's dyestuff activities until the end of the nineteenth century. It was not until Heinrich Brunck emerged as Engelhorn's successor in the mid-1880s, however, that BASF's era of "scientific competence, applications-oriented pragmatism, and longer-term planning" began in earnest and led to the inorganic chemical expertise key to the industrial upscaling of ammonia synthesis (Travis 1993a, pp. 163–178), (Abelshauser et al. 2004, pp. 23–37).

Originally, the inorganic department had served as BASF's own source of starting materials for the dye department to produce chlorine, anthranilic acid, and other chemicals for the lucrative indigo. However, helped by the pervasive spirit of innovation at the company, the inorganic department gained autonomy by the turn of the twentieth century through the advances and experience gained under the leadership of Rudolf Knietsch. The implementation of his contact process for the production of sulfuric acid, a precursor chemical for indigo, was so successful that BASF went from a consumer to net industry supplier. BASF's initial, failed attempts to synthesize ammonia from the elements with Wilhelm Ostwald in 1900 also took place in Knietsch's laboratories, as did research on the electric arc and cyanamide processes. During this time, the effect of pressure and different catalysts on chemical yields were examined and the influence of additives on catalytic materials became evident. These and other ventures helped BASF diversify beyond dyestuffs and were important for the later expansion of the inorganic chemical industry (Mittasch 1951, pp. 87–88), (von Nagel 1991, pp. 13–15), (Reinhardt 1997, p. 240), (Abelshauser et al. 2004, pp. 30–57, 70–72).

Other chemical companies were also founded in the 1860s, including Hoechst, Bayer, and Agfa. Coal tar dyes grew beyond their original fashionable niche into one of the most important products of the industry. The change was aided by a prominent showcase of the newly available dyestuffs at the 1862 International Exhibition in London. From humble beginnings with aniline, to the replacement of organic dyes with azo and synthetic alizarin dyes, to synthetic indigo, dyestuffs had developed significantly. By 1900, the sale of these products helped the chemical industry gain the financial means to fund ammonia synthesis research (Haber 1958, pp. 80–91), (Travis 1993a, pp. 214–228, 237–239), (Abelshauser et al. 2004, pp. 57–70).[8]

---

[8] The manufacture of explosives also played a part in the early chemical industry. Like dyestuffs, explosives are based on the nitrification of hydrocarbons such as phenol, resulting in picric acid. After mid-century the manufacturing process was based on scientific, chemical-technical knowledge using nitric acid and sulfuric acid. The former use of saltpeter alone had resulted only

As the nineteenth century drew to a close, the growth of the dye industry led to a more intertwined relationship with scientific chemical research (Basalla 1988, p. 28). The recognition of science as a profession had gained traction throughout the century and the increased involvement of academics as employees in the dye companies fostered a closer alliance. Whereas in 1800 there were about 1000 professional scientists in the world, that number grew to about 10,000 by 1850 and 100,000 by 1900 (Greenaway 1966, pp. 33–34). The occupation of the chemist was also effected by this trend. Whereas in 1800 there had been approximately 200 professional working chemists, the number increased to about 6000 by 1895 (Scholz 1987, p. 166). BASF itself had 78 chemists in 1892, 65 holding a PhD (Travis 1993a, p. 220). They were not only researchers; chemists were often in the highest positions in the industry and made crucial strategies and economic decisions.

Thus, at the beginning of the twentieth century, there was a strong tradition of science in Prussia awaiting Fritz Haber and Walther Nernst (Kennedy 1987, p. 209–211), (Schütt 1992, pp. 152–165), (Travis 1993a, pp. 62–64), (Leigh 2004, p. 113), (Harwood 2005, pp. 77–91), (Klein 2015a, pp. 17–20, 273–278). Through them "an old Berliner tradition of the 'relationships between chemistry and physics' found its continuation and consolidation in a distinguished way (Bartel 1989, p. 69)."[9] Or, if we turn to the words of Wilhelm Ostwald from 1903 (Ostwald 1903),

> If one asks about the means through which this state is reached, a decisive factor can be identified in the systematic exploitation of scientifically trained workers. No other nation has access to the human resources provided to German industry by our universities and trade schools. As long as this relationship is maintained any foreign competition may be discounted.[10]

At the end of the nineteenth century, the knowledge, infrastructure, and financial assets were largely in place to tackle the more challenging problem of ammonia synthesis from the elements (Szöllösi-Janze 1998b, pp. 161–162). As we have seen, other factors revealing the need for a synthesis process had been understood for some time. The integral role of ammonia and fixed nitrogen in agriculture had been accepted for decades, its importance underscored by the growing world population. The main source of this nitrogen, imports from the saltpeter mines in Chile, were a limited natural resource and subject to import blockades; they had also been the focus of the five year War of the Pacific. Global stability and national safety were

---

in a mechanical mixture called *blackpowder* (Kant 1983, pp. 23–28), (Akhavan 2004, chapter 7), (Aftalion 2001, pp. 53–57).

[9] Durch Nernst fand in hervorragender Weise eine alte Berliner Tradition ihre Fortführung und Vertiefung, die 'Beziehungen zwischen Chemie und Physik'. . ." The original quote refers only to Nernst, but I find it appropriate to include Haber.

[10] Fragt man sich nach den Mitteln, durch welche dieser Zustand erreicht ist, so läßt sich als entscheidender Faktor die systematische Verwertung wissenschaftlich geschulter Arbeitskräfte erkennen. Keine andere Nation hat das Menschenmaterial zur Verfügung, welches unsere Universitäten und technische Hochschulen der deutschen Industrie liefern, und so lange dies Verhältnis dauert, ist jede Konkurrenz des Auslandes ausgeschlossen.

in question. However, despite this context and decades of attempts to synthesize ammonia, a solution had not been found.

Here again we consider Alwin Mittasch's sentiments. The empirically based approach to chemical synthesis and catalysis in the nineteenth century made large contributions to the chemical industry, but they were not adequate to offer a solution to ammonia synthesis from the elements. There was still a missing component that would be found in a conceptual expansion within the scientific community itself. It allowed a holistic (or global) view and control of chemical reactions to alleviate the pressing problems of large scale, diverse industrial chemical production. This knowledge was gained during an intellectual transformation, during which scientists learned to view chemical processes as a problem at the intersection, or perhaps *defining* the intersection between chemistry and physics. Only with new theoretical tools was it possible to understand the fundamental concept of equilibrium in a chemical reaction and the exact role of a catalyst. The recognition, missed by the pioneers of catalysis, was that the dissociation and generation of a substance were linked and that the reaction was, therefore, *reversible*.

Chapter 4
The Mystery of Ammonia Synthesis

Even on the verge of Haber's breakthrough, scientists still stood in wonder of the behavior of nitrogen and its compounds. Quoting Ostwald from 1903 (Ostwald 1903),

> Among the chemical elements from which the body of living things, both the lowest and the highest, assembles itself, nitrogen plays a special, aristocratic role. While the other elements, oxygen, carbon, hydrogen, sulfur, iron, etc., are as willing to form chemical compounds as they are to dissociate from them, nitrogen forms compounds only with difficulty and is very inclined to leave them. An expression of this characteristic is the fact that free nitrogen makes up four-fifths of the atmosphere while fixed [nitrogen], which is mainly found in the solids and liquids of the Earth's crust, likely amounts to less than a millionth of it.[1]

The wonderment was not new (Sheppard 2020, p. 154).

By the end of the eighteenth century, ammonia was the subject of scientific inquiry and had been produced from other compounds. The substance itself had long been uniquely identified–it is readily detectable by its smell or "odeur très piquant"—but it was only after the chemical composition was determined that the idea of a synthesis from the elements could be considered (Mittasch 1951, pp. 1–24), (Timm 1984). Ammonia could be dissociated easily, but despite the simplicity of the apparently analogue reaction of water formation, the combination of H_2 and O_2, ammonia synthesis from the elements was subjected to a century of confusion and erroneous reports of success. The production of ammonia was repeatedly shown to have come not from direct combination of nitrogen and hydrogen but from the

---

[1] "Unter den chemischen Elementen, aus denen sich der Leib der Lebewesen, der niedrigsten wie der höchsten, zusammensetzt, spielt der Stickstoff eine besondere, aristokratische Rolle. Während die anderen Elemente, Sauerstoff, Kohlenstoff, Wasserstoff, Schwefel, Eisen u.s.w. eben so bereitwillig in chemische Verbindungen übergehen, wie sie sich aus ihnen wieder absondern, bildet der Stickstoff nur sehr schwierig Verbindungen und ist sehr geneigt, aus ihnen wieder auszutreten. Ein Ausdruck dieser Eigenschaft ist die Tatsache, daß der freie Stickstoff vier Fünftel der Atmosphäre ausmacht, während der gebundene, der sich meist in der festen und flüßigen Erdrinde befindet, wahrscheinlich weniger als ein Millionstel derselben beträgt."

© The Author(s) 2022
B. Johnson, *Making Ammonia*, https://doi.org/10.1007/978-3-030-85532-1_4

atmosphere, impurities in water, or other materials such as iron. The combination of the two elements must be an altogether different kind of reaction. The mystery was exacerbated by the pervasive presence of ammonia in nature, such as in the presence of rotting organic matter. What could nature do that man could not?

With our knowledge today, we can look back at the earliest experiments and know they were doomed to failure. Attempts to synthesize ammonia in the nineteenth century were repeatedly plagued by the same factors: temperatures and pressures were too low. Many believed direct synthesis was impossible (Haber 1920), a plausible position under the assumption that chemical reactions proceeded in only one direction. Before the fundamental ideas of equilibrium and reversibility were formulated, there could be no adequate understanding of the role of the catalyst. In cases where ammonia was produced with certainty from other compounds containing nitrogen and hydrogen, it was attributed to a particular state of matter called *status nascens*. Absolute proof of the direct combination of nitrogen with hydrogen could, however, not be found.

This is not to say that all experiments suffered from all of these deficiencies. By mid-century, there was regular use of higher temperatures, increased pressure, and catalysts, and in the last quarter of the nineteenth century elements of the emerging physical chemistry began to appear in experimental reports. Here our story converges to focus on these particular physical characteristics of ammonia production. It was the formation and dissociation of the ammonia molecule itself— the energetic nature and strength of its bonds and the bonds of its constituents, N_2 and H_2—when in contact with a catalyst that came under direct scrutiny. Despite the new scientific focus, however, the desperation of politicians, economists, scientists, and farmers to find a reliable source of fixed nitrogen remained.

As we have seen, knowledge of the nitrogen cycle had shown the element (in fixed form) to be the limiting factor in crop growth. With some success, the once closed nitrogen cycle between soil, crops, and manure had been opened through the use of fertilizer, transport of biomass, and drainage systems. A solution seemed feasible through further human control: scientists were intrigued by nature's ability to fix nitrogen biologically as well as through lightning and at the end of the nineteenth century imitation ensued (von Nagel 1991, pp. 9–15), (Stoltzenberg 1994, pp. 133–139).[2] While the attempts to directly breed and market nitrobacteria cultures failed, the multi-step, bacterial fixation mechanism was emulated in the cyanamide and other processes into the first quarter of the next century. But low yields could not be improved. Echoing the force of lightning, advances in electrical engineering made it possible to bind nitrogen through a new technique: the electric arc. Although the latter was also used well into the twentieth century, even cheap sources of electricity from hydropower in Scandinavia enabled only limited industrial and commercial application.

---

[2] Stoltzenberg's biography of Fritz Haber is also available in English (Stoltzenberg 2004). Page numbers may vary!

Then there was the direct synthesis of ammonia from the elements. The process, intriguing and elegant in theory, gained attention as the need for nitrogen fertilizer intensified. In practice, however, it remained unfeasible. Ammonia, while quickly dissociated catalytically, could not be produced in useful quantities because key physical insight was missing. All the while, a permanent solution for food production for the growing population had become a pressing topic in official circles. Crops were more and more constrained to specific geographical locations with increasing expectations on shrinking plots of arable land; there were also warnings that the nitrogen supplies in South America would soon be exhausted. In fact, these sources of fixed nitrogen, while having the potential to alleviate hunger, did more to better the lives of the wealthy while contributing to global instabilities, environmental destruction, and dismal working conditions; modern, high-throughput farming was now dependent on this stock resource (Cushman 2013, pp. 72–74). The limited quantities of ammonia from coking plants and other local supplies caused a permanent, synthetic source of fixed nitrogen to be viewed as a great necessity. In what has become a pivotal, historic appeal, the chemist William Crookes, also a student of August Wilhelm von Hofmann's, warned the British Association for the Advancement of Science in 1898 that the civilized nations were in danger of producing inadequate quantities of foodstuffs.[3] He pointed to the fixation of atmospheric nitrogen as a solution and his words also remind us of the mysterious nature of the process and its challenges (Crookes 1917, pp. 2–3),

My chief subject is of interest to the whole world–to every race–to every human being. It is of urgent importance to-day, and it is a life and death question for generations to come. I mean the question of Food supply. Many of my statements you may think are of the alarmist order; certainly they are depressing, but they are founded on stubborn facts. They show that England and all civilised nations stand in deadly peril of not having enough to eat. As mouths multiply, food resources dwindle. Land is a limited quantity, and the land that will grow wheat is absolutely dependent on difficult and capricious natural phenomena. I am constrained to show that our wheat-producing soil is totally unequal to the strain put upon it [. . .] It is the chemist who must come to the rescue of the threatened communities. It is through the laboratory that starvation may ultimately be turned into plenty.

He continued (Crookes 1917, pp. 37–38),

For years past attempts have been made to effect the fixation of atmospheric nitrogen, and some of the processes have met with sufficient partial success to warrant experimentalists in pushing their trials still further; but I think I am right in saying that no process has yet been brought to the notice of scientific or commercial men which can be considered successful either as regards cost or yield of product. It is possible, by several methods, to fix a certain amount of atmospheric nitrogen; but to the best of my knowledge no process has hitherto converted more than a small amount, and this at a cost largely in excess of the present market value of fixed nitrogen.

The fixation of atmospheric nitrogen therefore is one of the great discoveries awaiting the ingenuity of chemists [. . .] This unfulfilled problem, which so far has eluded the strenuous attempts of those who have tried to wrest the secret from nature, differs materially from other chemical discoveries, which are in the air so to speak, but are not yet matured.

---

[3] See also Frank (1903).

At the time of Crookes speech, nearly two decades before the First World War, the geopolitical dangers in the world were palpable. The Great Powers, Germany, France, and England were not only jostling for influence amongst each other, but also feared the world order that centered power on Europe was pivoting toward other nations: Russia, the United States, and Japan (Kennedy 1987, pp. 191–197), (Mackinder 1904; Wilkinson et al. 1904). Military and economic tensions were rising and "industrial productivity, with science and technology, became an ever more vital component of national strength (Kennedy 1987, p. 197)." Fixed nitrogen was used for many purposes besides fertilizer, including dyes and explosives for military, mining, and construction (Tamaru 1991). Not only were the resources limited, but the sea routes for importing nitrates from South America could easily be blocked. A new industrial method to produce fixed nitrogen would serve several purposes at once (Schwarte 1920, pp. 537–551), (Haber 1971, p. 85).

The final step, however, was not political. The industrial capital and infrastructure for chemical production was already in place. Fritz Haber would successfully synthesize ammonia only seven years later and the beginning of industrial upscaling was only a decade away. It was a development in pure science that offered the solution to the mystery of ammonia synthesis: the emergence of *physical chemistry*.

Chapter 5
Physical Chemistry: Uniting Two Branches of Science

The complexity of the chemistry that emerged out of the eighteenth century, especially chemical reactions containing organic compounds, quickly overwhelmed the ability of mechanical theories provided by physics to deal with a large number of bodies involved. These theories had worked well to describe the interactions of objects on our human scale as well as some celestial events, but failed to adequately clarify chemical phenomena. Despite the plausibility of speculation based on atomic principles, efforts to understand chemical processes on these grounds remained largely futile (Farber 1966, pp. 187–188). At the same time, the growing practical and industrial importance of chemical processes meant that an understanding of the physical parameters guiding a chemical reaction was gaining relevance.

It was the concept of the interconvertibility of heat and work which held the seed of the solution. Applied examples existed in the eighteenth century, such as the steam engine and the cannon (Malanima 2009, pp. 53–54, 63–65). While thermal machines made their mark on agriculture and other parts of society, it was the theoretical understanding of heat and work developed in the next century that allowed Rudolf Clausius, William Thomson, and William Rankine to publish the first formulations of thermodynamics in the 1850s. The first law was the conservation of energy: the sum of potential and "actual" (kinetic) energies in the universe remained constant.[1] It relied on the exact, mathematical formulation of mechanical work and its interconvertibility with kinetic energy (heat). The second law relied on Clausius' formulation of entropy. The application of these concepts to chemical reactions marked the birth of physical chemistry, enabling the creation of machines that make fertilizer (i.e. Haber-Bosch plants) (Kuhn 1959), (Farber 1966, pp. 150–156), (Girnus 1987), (Laidler 1993, pp. 1–11, 55–121), (Purrington 1993, pp. 80–101, 107–112), (Smith 1998, chapter 7), (Friedrich 2016).

Initially, the description of heat and energy successfully described the behavior of gases. The breakthrough in chemistry came in 1887 when the equation of state for

---

[1] The equations discussed in this section can be found in Appendices A and B using the formulations of Haber and Nernst.

© The Author(s) 2022
B. Johnson, *Making Ammonia*, https://doi.org/10.1007/978-3-030-85532-1_5

an ideal gas was expanded to describe chemical solutions (Windisch 1892, pp. 476–522), (Ostwald 1927, pp. 16–31), (Partington 1964, pp. 637–662, 663–681). Jacobus Henricus van't Hoff and Svante Arrhenius were able to equate the pressure of a gas to the osmotic pressure in a solution so that the ideal gas law could be applied in both cases (Arrhenius 1887; van't Hoff 1887). While this discovery built the foundation of physical chemistry (Ostwald 1927, p. 22), it was another more general thermodynamic concept that was central to Haber's and Nernst's work: the free energy.

Until the 1870s, the spontaneity and direction of a chemical reaction was thought to be determined by the *principle of maximum work* which stated that the specific reaction that occured was the one that released the most heat (and could thus produce the most work) (Nernst 1914), (Partington 1964, pp. 610–620, 684–699), (Farber 1966, pp. 156–157), (Suhling 1972), (Bartel 1989, pp. 71–73), (Smith 1998, pp. 260–263). However, there was evidence, such as endothermic reactions, which contradicted this hypothesis. It was, in fact, a special kind of energy that dictated the behavior of a reaction: the free energy. This was envisioned as an "ordered" energy within the system available to perform work. When the free energy was at a minimum, the reaction stopped and the system was in equilibrium.[2] There was also a form of "disordered" or "bound" energy, which only manifested itself as heat and did not dictate the course or direction of the reaction (at least at finite temperatures). The total heat released in a reaction was a combination of these two types of energy. Josiah Willard Gibbs and Hermann von Helmholtz independently formulated expressions for the free energy in the 1870s in terms of the state functions of a system: temperature, pressure, volume, entropy, and total energy (Gibbs 1878a,b; von Helmholtz 1882).[3] With their theories, quantitative predictions could be made about chemical reactions in terms of global variables, which could be represented within a "free energy landscape." Gibbs also considered pure and mixed-phase systems for which he described the chemical potential, or free energy per species in the reaction.

In terms of the actual application of these ideas to experiment, there was a further necessary development in the 1880s linking the free energy to chemical equilibrium. van't Hoff, not employing the mathematical rigor of Gibbs or Helmholtz, derived a practical, yet theoretical description of chemical dynamics (Fig. 5.1). The most important quantity was the equilibrium constant which determines how much of each substance is present at equilibrium under specific conditions, which van't Hoff expressed in terms of the state variables temperature and free energy. The idea of an equilibrium constant had been developed in the 1860s, but it was not until van't Hoff's ideas were understood that the dynamic nature of chemical equilibrium

---

[2] Helmholtz directly described a minimum in the free energy, F, in 1882 with $\Delta F = 0$. Haber, in his practical derivations of chemical equilibrium, used thermal data to calculate the equilibrium expected from a "standard" free energy change during a chemical reaction and set his measurement of equilibrium equal to the calculated value (see Appendix A).

[3] These were not the first formulations of the free energy. See for example (Partington 1964, pp. 614–616).

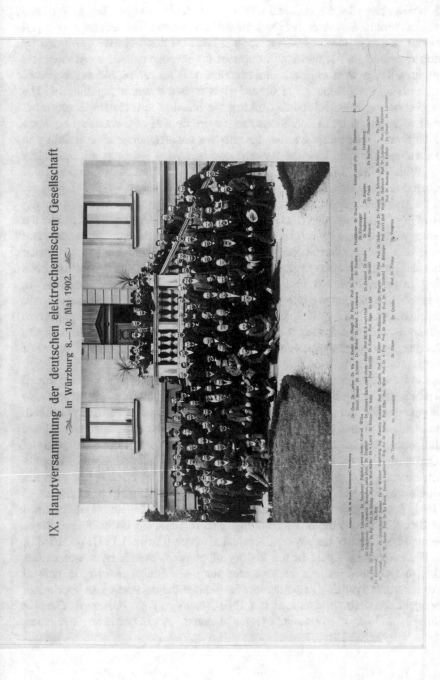

Fig. 5.1 9th Meeting of the German Electrochemical Society in Würzburg, May 8–10, 1902. Several of the architects of the theory of physical chemistry discussed here are pictured. Second Row: Walther Nernst, far left; Wilhelm Ostwald, middle with head turned; Jacobus Henricus van't Hoff, two to Ostwald's left; Third row: Guido Bodländer, middle, behind and to Ostwald's right, Fritz Haber, three to Bodländer's left. Back row: Heinrich Danneel, behind and just to Haber's left. *Source*: Archiv der Berlin-Brandenburgischen Akademie der Wissenschaften, NL Ostwald Nr. 5296

could be quantitatively applied to experiment. It had been thought that a chemical reaction was driven by the classical notion of force due to chemical affinity and that at equilibrium the forces for the forward and backward reactions were equal and opposite–a very mechanical approach. Reactions are, however, governed by reaction rates and do not find equilibrium in the mechanical sense. At chemical equilibrium, the forward and backward reaction rates for each substance are equal and there is no net production or dissociation; it is a dynamic equilibrium. The production of a specific substance (driving the reaction in a certain direction) is a matter of finding a set of conditions that favor one side of a chemical equation in equilibrium. This concept clarifies the modern understanding of reversibility in a chemical reaction and how the forward and backward reactions are intimately linked. With this revelation, the role and effect of a catalyst as an accelerator of these reaction rates could be understood.

At the end of the 1880s, the theoretical structure of physical chemistry, having emerged from thermodynamics, included reversibility, dynamic equilibrium, and the free energy. These were consolidated in the expression for the equilibrium constant and enabled a global understanding of a chemical reaction. This was the step beyond mechanical models and the empirical approach of the past century that had been missing. In the case of ammonia, the existence of even the smallest quantities in the presence of hydrogen and nitrogen meant there was a chemical equilibrium, and that ammonia was being actively produced (and dissociated). Once the equilibrium constant was determined, one could identify values of the state functions of the system (temperature, pressure, volume, energy) that would raise the reaction rate of ammonia production.

Due to these developments, the year 1887 can be considered the point at which physical chemistry became a distinct discipline (Ostwald 1927, p. 19). It was also in this year that the journal *Zeitschrift für Physikalische Chemie* was first published. For ammonia synthesis, only two problems remained at the time: one practical and one theoretical. The solution to the practical problem was the work of Haber, Gabriel van Oordt, and Robert Le Rossignol, and later Carl Bosch and Alwin Mittasch. The theoretical problem was tackled by Walter Nernst, who came out of Wilhelm Ostwald's renowned Institute for Physical Chemistry in Leipzig. Reminiscent of the Liebig School, generations of internationally recognized chemists were trained under Ostwald; at the time of its founding, also in 1887, his institute was unique in the world as a center for physicochemical research (Ostwald 1927, pp. 16–31). Himself an important figure in the history of physical chemistry, it was at this institute that Ostwald made a decisive step for the chemical industry in general and for ammonia synthesis in particular: he defined the catalyst as an accelerator of chemical reactions. In a talk at the 1901 *Meeting of the Society of German Scientists and Doctors*[4] in Hamburg Ostwald stated, "A catalyst is any substance,

---

[4] *Versammlung der Gesellschaft Deutscher Naturforscher und Ärzte.*

that, without having an effect on the end product of a chemical reaction, changes its speed [reaction rates] (Ostwald 1902)."[5]

Several years after Ostwald's initial definition, the role of a catalyst as a promotor was easily measurable. Chemical reactions could be systematically accelerated and became more profitable. The transformation of resources and the transformation of an entire scientific field had become inextricably linked with the transformation of an industry. The chemical industry, at this point consisting largely of the dye manufacturers, was dependent on catalysts for certain processes. An example is the aforementioned contact process that BASF implemented in the late 1890s to produce sulfuric acid. It enabled the production of heretofore unobtainable quantities of acid and, along with synthetic indigo and ammonia synthesis, came to symbolize the birth of the modern chemical industry (Stranges 2000, p. 170), (Marsch 2000, pp. 220–223).

A catalyst, however, could not change the end result of a chemical reaction; only the reaction rates could be manipulated. The free energy was still vital to understanding the direction of a reaction and how far it would proceed. The question was whether the value of the free energy could be calculated rather than determined through a laborious and expensive experimental observation. As mentioned, the free energy is a measure of the maximum amount of work (useful energy) that can be obtained from a chemical system. The spontaneous reaction is the one which minimizes the free energy as it is converted into work during a chemical reaction; at the minimum, the system is in chemical equilibrium. For a prediction or calculation of the free energy a complex, temperature-dependent function of total energy, free energy, and entropy is needed (Müller 2007, p. 148).[6] Mathematically speaking, one has to solve a differential equation, a difficulty that became one of the central theoretical challenges of physical chemistry at the end of the nineteenth century (Haberditzl 1960; Suhling 1972). In order to obtain a temperature-dependent expression for the free energy, the differential equation must be integrated, a mathematical operation that results in an unknown constant of integration. That means the form of the curve for the change of free energy with temperature was known, but its absolute value was not. In 1900, it was customary to solve this problem by using an empirical measurement of equilibrium quantities to determine the constant. Although this method worked in principle, it burdened an otherwise exact theory with experimental error (Nernst 1914).[7]

Walther Nernst, who habilitated under Wilhelm Ostwald in Leipzig, finally found a solution to this fundamental problem. Nernst was among the first generation of

---

[5] "Ein Katalysator ist jeder Stoff, der, ohne im Endprodukt einer chemischen Reaktion zu erscheinen, ihre Geschwindigkeit verändert."

[6] In terms of the previously mentioned "free" (ordered) and "bound" (disordered) energies, the useful "ordered" energy in a chemical system is equal to the total energy minus the "disordered" energy. The latter is equal to the temperature multiplied by the change in entropy. A higher temperature or entropy change means an increase in "disordered" energy and a corresponding loss of "ordered" energy.

[7] See also Haber and van Oordt's methods from 1905 (Haber and van Oordt 1905b).

chemists for whom the career choice was considered "acceptable"—such was the state of approval toward scientific professions. Like Liebig, Nernst was a modern chemist, who advocated for basic scientific principles and mathematics in his field, meaning he was (and often still is) considered more of a physicist. In 1887, he joined Ostwald in Leipzig where he quickly became involved in the investigations of the heat developed during chemical reactions and their relationship to chemical equilibria. In 1891, Nernst moved to Göttingen where he became professor in 1894 and head of the new Institute of Physical Chemistry and Electrochemistry the following year. There, his thermochemical studies continued and laid the groundwork for his seminal theoretical contribution to physical chemistry in 1906 when he published his heat theorem in "Ueber die Berechnung chemischer Gleichgewichte aus thermischen Messungen" (On the Calculation of Chemical Equilibria from Thermal Measurements) at the University of Berlin (Nernst 1906), (Bartel 1989, pp. 10–45). The theory allowed Nernst to calculate an expression for the unknown integration constant by introducing a mathematical formulation of the third law of thermodynamics. Neither Nernst nor anyone else made much of the physical significance of his postulate at the time. The third law states that the change in the free energy and of the total energy of a system with temperature will become identical at the absolute zero of temperature. In fact, the change in both energies itself also tends to zero. The new equation enabled Nernst to calculate the integration constant from measurements of heat capacities of the chemicals involved in a reaction (see Appendix B).[8] The determination of the integration constant became independent of any actual scientific observation of the reaction itself. It depended only on basic characteristics of the pure substances involved which could be measured more simply and exactly than tiny amounts of a chemical yield in equilibrium. Nernst's method was not only an advancement for academic chemistry, but also for industry as it became more economical to select suitable chemical reactions.

In view of the development of physical chemistry, the problem of ammonia production on the basis of the necessary physicochemical principles could only have been taken up in the 1890s. Although an initial attempt to industrialize ammonia production by Ostwald failed in 1900, it paved the way for the experimental aptitude of Haber and the theoretical insight of Nernst in the new century. The new thermodynamic descriptions were like lines of latitude and longitude on a map—they set up a closely spaced grid to describe the successive steps of a chemical reaction. The state variables defined specific coordinates on the grid and chemical processes could be followed with a precision that ushered in a new era of chemistry and chemical production.

---

[8] Alternatively, if the heat of reaction was known from thermal measurements, Nernst's equation allowed the absolute calculation of the free energy.

Chapter 6
The Scientific Breakthrough (1903–1908)

So far, we have discussed a confluence of factors leading to the ability to understand and solve the practical problem of ammonia synthesis. Research in agricultural science led to an awareness of the need for a synthetic source of fixed nitrogen, while developments in organic chemistry helped lead to the infrastructure, capital, and expertise of the German chemical industry. Further conceptual steps in the natural sciences gave rise to the discipline of physical chemistry and a new theoretical framework for chemical reactions. The possibility of synthesizing ammonia from the elements was now plausible, meaning a solution was on the horizon. In 1898, William Crookes' message to the world on the importance of ammonia synthesis for the production of foodstuffs not only framed the looming humanitarian crisis but also made its economic potential clear. All that was left to determine were the conditions under which ammonia could be synthesized.

In the century before Haber's work, scientists had taken up the problem of ammonia synthesis and although several patents had been taken out, none were successful (Mittasch 1951; Tamaru 1991). One of the first chemists to undertake the challenge using a physicochemical approach was Wilhelm Ostwald in the 1890s. His experiments were conducted at atmospheric pressure with temperatures between 250 °C and 300 °C. He also used a catalyst to raise the reaction rate. Any ammonia formed was removed from the continuously circulating reaction gas mixture in a cycle that was also responsible for heat exchange. Ammonia formed at higher temperatures and was extracted at lower temperatures while the excess heat was used to warm incoming gas precursors. Today, we recognize that Ostwald's methods were, in principle, correct. He did not, however, possess the necessary experimental sensitivity with which a measurement of ammonia equilibrium must be made, nor was he mindful enough of the impact of impure materials. What he thought was a successful synthesis of ammonia from the elements was actually the result of a contaminated experiment: the ammonia detected had been produced from nitrogen in the iron catalyst. It was a young Carl Bosch who, while attempting to reproduce Ostwald's results at BASF, uncovered the error and dashed Ostwald's

© The Author(s) 2022
B. Johnson, *Making Ammonia*, https://doi.org/10.1007/978-3-030-85532-1_6

dreams of a lucrative patent for ammonia production (Ostwald 1927, pp. 279–287), (Farbwerke Hoechst 1964), (Holdermann 1960, pp. 40–42).

While Ostwald moved on to different endeavors, work on ammonia synthesis from the elements continued. Reacting to a proposal by the Margulies Brothers from the Austrian Chemical Works in 1903, the physical chemist Fritz Haber began working on the ammonia system (Haber 1903c). Using what amounted to Ostwald's laboratory method, Haber worked at even higher temperatures and tested different catalysts to further raise reaction rates. In 1905, he published his results (Haber and van Oordt 1905b). At the extreme temperature of 1023°C he found only 0.012% ammonia in equilibrium with nitrogen and hydrogen. His experiments were successful, but his findings did not support the possibility of an economically viable synthesis as the engineering challenges appeared too great. Although working at higher pressures would have brought Haber more reliable results, he at first decided against the strategy due to technical difficulties. Haber's publication led to a famous written response from Walther Nernst in 1906 and an exchange at the 14th Meeting of the Bunsen Society[1] in 1907 in Hamburg during which Nernst drew attention to "Haber's very inaccurate numbers."[2] They were, in Nernst's view, not in accordance with his heat theorem (Nernst 1906, 1907).

Nernst proposed Haber repeat the measurements under higher pressures to increase ammonia yield and improve measurement accuracy. Haber knew this concept well, but only after the motivation of a senior scientist did he make use of it. For his part, Haber (correctly) cast doubt on Nernst's methodology. The exchange is often characterized as a polemic between two self-absorbed scientists, but it was, in fact, an important event in the development of ammonia synthesis. The interaction between Haber and Nernst enabled a complementary exchange of knowledge and suggestions that guided Haber toward the solution of the ammonia mystery. The key was that they approached the synthesis of ammonia from different perspectives. Haber, experienced in the laboratory, was aware of the required measurement sensitivity as well as the importance of using pure substances. His experimental aptitude was also put to the test when determining equilibrium conditions in the laboratory "from both sides"[3] with the necessary accuracy. Nernst, ever the theorist, was known for his critical nature and hence focused on how the experiment could be improved on theoretical grounds. In this case, he drew on the physicochemical concept that higher pressures would increase the amount of ammonia present in equilibrium. He also wished to avoid fixing the integration constant for the free energy with an experimental result and so referred self-confidently to his theoretical work. He saw ammonia synthesis as a test for his conception of physical chemistry: theory and experiment must complement one another and ultimately arrive at the same result. We can see explicitly from these approaches how their interaction

---

[1] 14. Hauptversammlung der Bunsengesellschaft.

[2] "...die stark unrichtigen Zahlen Habers...".

[3] This was the expression used to denote experiments involving both the production and dissociation of ammonia.

bridged the gap between experiment and theory. The roles of Haber and Nernst were not restricted to that of the experimentalist and the theorist, however—neither was purely one nor the other—and both knew it. Their experience and abilities assured their interaction would be an exchange of views rather than a lesson from one to the other. Haber, for example, had successfully used entropy in his theoretical thermodynamic derivations while Nernst had an aversion to the quantity. In his initial formulation of the third law, Nernst chose not to mention entropy although one of the quantities he used was equal to the entropy of the system. On the other hand, Nernst had experience in experimental methods reaching back two decades. He saw little difference between pure and applied research (Bartel 1989; Haberditzl 1960; Suhling 1972). Both men understood that the synthesis of ammonia was in principle a balancing act. High temperatures increased the reaction rate, but above a certain temperature, the catalyst became inactive and the decomposition of ammonia began to outpace production. Basic physicochemical principles were not in question.

The scientific breakthrough was only possible through a cooperative effort. As Rudwick puts it in the case of The Great Devonian Controversy, the outcome "was clearly *not* established by the straightforward victory of one interpretation or theory over its rival or competitor (Rudwick 1985, p. 405)." Haber began with Ostwald's experimental plan and benefited from Nernst's suggestions. Nernst's criticism and insistence on theoretical stringency was the reason Haber was able to see more in his measurements than merely an approximate determination of the integration constant for the free energy. His comments drove Haber toward more fundamental questions. Haber himself can be credited with recognizing the need for increased measurement sensitivity and pure substances, and for his and his assistants' ability to perform the experiments with high accuracy. They also recognized the effect of an inferior catalyst. In the end, the exchange at the meeting of the Bunsen Society was, along with Haber and Nernst's other public and private exchanges, vital and fundamental to the scientific breakthrough. It was a dispute among scientists with beneficial results. A similar dynamic still plays out today.

The path to ammonia synthesis in a scientific laboratory setting had ended, but still unresolved was the technical engineering problem of bridging the divide between scientific understanding and an economically viable manufacturing process. It was unclear if it was possible at all.

Chapter 7
The Challenge of Technical Implementation

In 1908, Fritz Haber entered into a contractual agreement with BASF to develop, on the basis of his discovery, an industrial-scale, high pressure synthesis of ammonia (Haber 1971, pp. 92–97), (Stoltzenberg 1994, pp. 144–170), (Szöllösi-Janze 1998b, pp. 171–181). BASF was no stranger to daunting challenges: the production of indigo had taken a full 17 years to develop. In addition, there was now Nernst's theoretical description of ammonia synthesis which assured the company that Haber's determination of the equilibrium of ammonia was correct (Haberditzl 1960; Suhling 1972). The bet paid off. On July 2, 1909, Haber and Le Rossignol delivered a working laboratory scale apparatus for the continuous production of ammonia. At that point, two main hurdles separated Haber and BASF from an industrial solution. First, a completely new high pressure system was needed to withstand enormous working pressures and temperatures. Second, suitable catalysts needed to be found that were both economical and effective. The experimentally successful osmium from the incandescent light bulb industry provided a laboratory scale solution, but the worldwide supply amounted to only several dozen kilograms, making it, like the costly uranium, a poor fit for industrial production.

The efforts to overcome the engineering challenges were led by Carl Bosch (Bosch 1932), (von Nagel 1991, pp. 26–33), (Holdermann 1960, pp. 65–102). The extreme conditions for the continual production of ammonia required constant reassessment of the use of raw materials and energy. Looking back in 1933, Bosch succinctly stated the demands of large scale ammonia synthesis (Bosch 1933),

> It is not enough to master the process in the laboratory. Rather, the production of hydrogen and the appropriate design of the reaction chambers must be solved in a technical and economical manner along with the viability and stability of the catalysts and other materials.[1]

---

[1] "Es genügt nicht, den Prozeß im Laboratorium zu beherrschen, sondern es müssen in technischer und wirtschaftlicher Hinsicht die Herstellung des Wasserstoffs, die geeignete Ausgestaltung der Reaktionsräume und die Brauchbarkeit und Haltbarkeit der Katalysatoren sowie der Materialien gelöst werden."

© The Author(s) 2022
B. Johnson, *Making Ammonia*, https://doi.org/10.1007/978-3-030-85532-1_7

One of the main difficulties was to produce a large enough quantity of the nitrogen-hydrogen gas mixture that was to be converted into ammonia. Nitrogen could be produced in a straightforward way via air separation, but the production and purity of hydrogen had a larger effect on the synthesis of ammonia. In the end, several different methods were employed, the most suitable of which was the catalytic conversion of carbon monoxide and steam to carbon dioxide and hydrogen based on the water gas shift reaction, but with a novel iron oxide-chromium oxide catalyst (von Nagel 1991, pp. 34–37). Hydrogen was to be produced from air, coal, coke, and water in "the largest gas production facilities ever built (Bosch 1933)."[2] They would eventually reach rates of $\sim$1 Million m^3/hr. A corresponding quantity of nitrogen needed to be produced and, along with the hydrogen, cleaned of carbon monoxide, hydrogen sulfide (which poisoned the catalysts), and other contaminants. These processes ran at the unprecedented pressure of 200 atmospheres and proper materials were needed to withstand the resulting forces and high heat while allowing thermal conduction to regulate temperature.

The challenges for the synthesis of ammonia itself proved even more severe. Bosch, starting from Haber's laboratory apparatus, designed reaction tubes to withstand longer production times. The first tubes, however, burst after only about 80 hours of use. The walls of the tubes were made of carbon steel (carbonaceous perlite dispersed in pure iron) and the hydrogen inside the reaction space had defused into the walls where decarbonization of the perlite led to the formation of methane. The result was a brittle alloy (iron hydride). The working conditions under high pressure and above 400 °C meant that degradation was inevitable after hours or days, no matter the material used (of those available at the time). Bosch's proposed solution was to separate the converter into two functional parts: one to withstand the high working pressures and one to provide a gas-tight seal. The special construction could be achieved by a converter consisting of a thin, soft steal inner lining, surrounded by a pressure-bearing steel mantle with perforations (Bosch holes). The hydrogen could diffuse through the soft steel (which would become brittle but remain intact) and escape through the holes in the mantle without compromising its structural integrity. The inner lining could be replaced as needed. Another challenge was the temperature. The hydrogenation processes had to be held within narrow temperature limits, making temperature control essential. To avoid excessive temperature gradients and conserve energy, a heat exchange system was developed, similar to those used by Ostwald and Haber, that took heat from the exothermic chemical reactions and warmed the incoming gases. The functionality of the system was vital: if the reaction drifted outside of the temperature limits, secondary reactions could release more heat, leading to a chain reaction and drastic material damage (Bosch 1932, pp. 200–221), (von Nagel 1991, pp. 26–33). Bosch was able to overcoming these and other hurdles and began the industrial scale production of hydrogen and ammonia (Fig. 7.1). The experience later helped him with the hydrogenation of carbon monoxide to methanol, and of coal, crude oil, and

---

[2] . . .die größten Gaserzeugungseinheiten, die bislang gebaut worden sind."

Fig. 7.1 This page: installation of a high-pressure contact reactor in 1920. Opposite page: A high-pressure contact oven in 1943. Comparison with Haber and Le Rossignol's laboratory apparatus in Fig. 12.1 of Part II, Chap. 12 illustrates the challenge of sheer magnitude faced by Carl Bosch and Alwin Mittasch. *Source*: BASF SE, Corporate History, Ludwigshafen

Fig. 7.1 (continued)

tar to gasoline—processes that became the backbone of the modern high-pressure catalytic industry.

Of equal importance were the catalysts for hydrogen production and for ammonia synthesis. Bosch followed these developments closely as the production facilities were planned in tandem with catalyst research; the efficacy of the catalyst determined the working temperature and pressure as well as the volume of the reaction chambers. The catalyst research itself was carried out in a separate department of BASF under the leadership of Alwin Mittasch, also a student of Ostwald's. He planned and executed a gigantic experimental program to investigate and test thousands of different contact materials (Mittasch 1951, pp. 91–121), (von Nagel 1991, pp. 23–25). BASF had the added benefit of extensive experience in sulfuric acid production (Knietsch's contact process) and in ammonia synthesis via the multi-step process (barium cyanide and titanium nitride), and through direct synthesis during the previous cooperation with Ostwald. This background pertained not only to catalyst materials, but also to specific knowledge about additives and catalyst poisoning. In only a few years, over 20,000 materials and material mixtures were tested on an industrial scale, including iron and other elements in the iron group such as osmium, uranium, manganese, cobalt, and nickel—an inconceivable undertaking in an academic setting. Their suitability was assessed on the basis of physical structure as well as on the ability of additives to activate or poison the catalyst. Using up to 20 high pressure ovens at a time, the BASF ammonia laboratory could test 50 to 100 different materials every week. Soon the result was clear: a highly porous iron catalyst with alumina and alkali additives was the best compromise between efficacy and cost. The interweaving of laboratory structure, financial assets of the industry, and academia produced the innovation. Physical chemistry played an important role, but not during every practical step. Modern scientific concepts were employed in tandem with the centuries-old strategy of trial and error. Mittasch, as had been done in the 1800s, made skilled use of the empirical method. There was no way to find appropriate catalysts except to manufacture and test each possibility.

Fritz Haber also reminds us of the importance of experience and experimental courage in opening new avenues of research. He and Mittasch had placed great importance on the use of pure substances in the laboratory. In a letter to the board of BASF in 1910, Haber wrote (Haber 1910d),

> Iron, first used by Ostwald and which we also tried hundreds of times in a pure form, finally worked in an impure state.[3]

The impurity was, of course, targeted rather than random. Mittasch's insight into the role of additives, activation, and poisoning as well as the physical structure of catalysts went further than the development of high pressure synthesis itself. With Mittasch, catalytic chemistry became a technical tool, especially in the later application of the power of mixed-material catalysts to the field of hydrocarbon chemistry. This progress led to the emergence of petroleum chemistry and the formation of the material basis of the twentieth century. The molecule became the building block of the chemical industry and could be paired in any combination with a particular catalyst to achieve the desired result. However, the development of catalytic chemistry was not self-evident as BASF opened the first ammonia synthesis plant in Oppau in 1913. Just one year later the First World War began and threw the nitrogen market into a tumult. Suddenly, the nitrogen sources in South America were no longer accessible to the Central Powers and the only alternative was to boost internal industrial capacities. In Germany, giant ammonia production facilities were built at Leuna. Haber, Bosch, and Mittasch's synthesis method, which a short time before had been insignificant, was propelled into the forefront of a worldwide industry. The necessities of war changed Germany from a nitrogen importing nation into a net exporter (Stoltzenberg 1994, pp. 186–189), (Szöllösi-Janze 1998b, pp. 185–191, 274–316).

Paradoxically, even as Germany was losing World War I, the Haber-Bosch-Mittasch process decisively brought its local nitrogen industry to the world market. After the end of the war in 1919, price controls for nitrogen had to be implemented for the protection of the agricultural industry because modern farming methods

---

[3] "Also das Eisen, mit dem Ostwald zuerst gearbeitet hat, das wir dann hundertfältig in einem reinen Zustand probiert haben, wirkt nun im unreinen Zustand."

had grown entirely dependent on synthetic sources of fertilizer (Schwarte 1920, pp. 542–551), (Szöllösi-Janze 2000, pp. 91–121). Such was the unprecedented control humans had gained over the nitrogen cycle.

The long march across two centuries toward agricultural optimization came to a close in the early 1920s. Or was it more of an overshoot?

Chapter 8
Reflections on Scientific Discovery and *The Haze*

The presentation of the confluence of factors has often included a modern perspective. Here I will touch on some consequences of Haber's breakthrough before reviewing aspects important to the next two parts of the book.

First, the current environmental and ecological repercussions of the scientific and technological breakthrough for feeding the world's population and the manufacturing of explosives were already appearing in the first half of the twentieth century as the possibility to mass-produce fixed nitrogen in Haber-Bosch (and other) facilities spread across the planet (Travis 2018, chapters 12–15). Although we can now produce enough food to feed the world, the nitrogen shortage has become an excess that threatens water, air, and soil quality as well as entire ecosystems and biodiversity (Ertl and Soentgen 2015; Sutton et al. 2011c). Our arable land is saturated with nitrates, which are transported through runoff into environments, especially coastal regions and oceans, not intended to receive such chemical surpluses. The resulting imbalance impacts life in these areas, often adversely through eutrophication and hypoxia. It is here, at the interface of the terrestrial and marine nitrogen cycles that the dynamic is most complex. Total synthetic fixed nitrogen production is now at the same order of magnitude as natural production either on land or in the ocean ($\sim 10^2$ Tg/year). In combination with natural conditions, it has contributed to changing quantities of biomass and species composition (Voss et al. 2011).

The effects are not only found in the natural environment, but also in our urban centers and in fundamental changes in our society. The production of food is no longer the mere collection of the annual influx of solar energy, it now consumes other sources of power as well. In achieving this change, we have used technical innovation along with energy and raw materials from fossil fuels to boost a large input mechanism into the nitrogen cycle: fertilizer. Advances underlying modern mobility have also resulted in new input mechanisms via the internal combustion engine. Driving and other forms of combustion (which we in the modern world have strategically hidden out of sight to the extent that we forget about their existence) have not only led to smog in our cities but also to noticeable growth of lichen and

© The Author(s) 2022
B. Johnson, *Making Ammonia*, https://doi.org/10.1007/978-3-030-85532-1_8

moss along heavily used traffic routes (Soentgen and Cyrys 2015).[1] The nitrogen cycle has not only been opened; man-made factors have bound the carbon and nitrogen cycles together in an unprecedented way. Hydrogen for the Haber-Bosch process is obtained by stripping it from hydrocarbons, and the residual nitrogen runoff effects the growth of biomass, linking it to levels of carbon, phosphorous, and other elements in biogeochemical cycles (Gruber and Galloway 2008; Sterner et al. 2008). Among other consequences, the effect of certain levels of greenhouse gases and the future of our climate have become difficult to assess, making fixed-nitrogen production one of the most critical developments of our time.

This rather dire prognosis is not meant to support the conclusion that the consequences of synthetic fixed nitrogen are inordinately or inherently negative, but rather to remind us of the risks of their extensive use. We have already changed and damaged our natural environment with the output from the power-hungry Haber-Bosch process, which consumes 1–2% of the world's total energy production. We need to find a better way to meet our needs. Possible solutions include forms of fertilizer that release fixed nitrogen in more efficient ways, improvements in existing technologies, and new methods of synthesis (Douat et al. 2016; Erisman et al. 2008; Hawtof et al. 2019; Patil et al. 2018; Rafiqul et al. 2008; Schrock 2006; Wissemeier 2015). A provocative detail of the latter is that they include plasma, or electric arc technology, which was originally the favored production method of Fritz Haber and most of the physical chemists at the turn of twentieth century (as we will see in Part II). We may, therefore, still come full circle, but this time we will be equipped with a wider range of options than every before. Technology alone will not solve all our problems, however. An integrated approach is essential, which spans science, technology, industry, politics, and of special importance, voices from the public sphere (Bull et al. 2011; BUND 2012; Reay et al. 2011; Sutton et al. 2011a).

While the environmental impact is vast, it was certainly not one of Haber's objectives, nor should we blame him for it. Scientists do not, without further involvement, bare responsibility for the technological repercussions of their discoveries. However, moral questions do crop up with respect to another aspect of ammonia synthesis.

The production of fertilizer and the manufacture of explosives in Europe in the first decade of the twentieth century was heavily dependent on nitrate imports. As the First World War loomed, the possibility, and eventual actuality, of blocked maritime trading routes caused many in Germany to reconsider the general security of the country and led to massive investment in science and technology to obviate possible shortfalls. Beyond a synthetic source of nitrates, much of Haber's pre-war work investigating gas reactions also leant itself to another possibility for the war effort: the development of poison gas. This new type of weapon promised not only a solution to munitions shortages but also enabled a new strategy of warfare (Haber 1971, pp. 208–217), (Friedrich and James 2017a; Szöllösi-Janze

---

[1] As an example, ammonium nitrate, an excellent fertilizer, is formed when ammonia from a diesel catalytic converter combines with nitrogen dioxide.

2017). Many scientists, some very prominent, became involved in research and deployment of chemical weapons, but it was Haber who notoriously proposed the use of chlorine gas. He was so central to the effort that one scientist even defined the product of gas concentration and exposure time before a subject's death as the "Haber Constant." After initial lackluster tests and attempted deployment in the beginning of 1915, chlorine gas was used at Ypres on April 22 against French and British troops; the cloud of poisonous gas spread over the battle field, killing hundreds and injuring thousands. Within months Haber, already heavily involved in the German mobilization effort, redirected research at his Kaiser Wilhelm Institute for Physical Chemistry and Electrochemistry toward gas warfare. Besides chlorine, research was conducted on mustard gas, phosgene, and other aggressive compounds as well as gas masks and respirators. Both offensive and defensive solutions were considered as the Entente Powers began to develop their own chemical weapons. Haber's institute grew rapidly in size. Not only personnel, but also the number of buildings and research locations on-site, in Berlin, and around Germany expanded. Haber relished his role and influence as one of the earliest "intermediary experts" as the convergence of state, military, economy, and science began to transform research into something like what we recognize today as big science. This development also continued to broaden the scope of the professional chemist which had started in the early nineteenth century (Fig. 8.1).

Fig. 8.1 Haber, second from left, inspecting gas munitions during the First World War. *Source*: Archive of the Max Planck Society, Berlin-Dahlem, Jaenicke Collection, Picture Number IX/9

The use of poisonous gas (by belligerent parties on both sides) went against international law and garnered for Haber, in addition to general moral critique, accusations of war crimes. While Haber had insisted on the "humane" nature of chemical weapons to shorten the length of the war and limit the number of deaths, this moral stance apparently counted only in the case of German victory. He also saw it as a way to weed out unfit soldiers. Later, Haber admitted no new science had resulted from chemical weapons research (although technological progress was made) but never regretted his involvement in chemical warfare–blame for wrongdoing was ultimately placed at the feet of others. During the war years, Haber committed his entire scientific effort to this work and continued it into the interwar period in clandestine fashion under the guise of extermination of vermin and pests (the claim was not wholly without legitimacy). During this research, scientists determined methods for safely working with cyanides, one of which was called Zyklon A. Unforeseen by Haber, it's successor, Zyklon B, was later used against civilian populations. Research on poison gas, much like ammonia synthesis, shows the double-edged sword embodied by scientific research, and in many ways by the person of Fritz Haber (2019).

Observing the entirety of these events, it is clear that pure science and technological breakthroughs lead to changes in systems of knowledge, means, and resources. Here, I have described one pathway leading to success (along with the corresponding consequences) that illuminates how a collection of phenomena spanning 150 years can be unified into a single framework. It was an arduous development and represents the main barrier to scientific progress: the inherently slow pace of interlinking of knowledge and processes, and the "scientification" of basic principles in a new field. While it may not have been the only possibility, I have tried to elucidate the basic pathway of advancement that, in this case, led to the final discovery.

The term "discovery" returns us to the discussion of a flow of a events, demarcated in a deliberate and informative way. Which set of events contains the knowledge that is awarded the label "discovery"? The declaration of *when* something was known is tied to *who* discovered it and which supporting evidence was used. It is also very sensitive to how the discovery itself is defined. The *who* has also been discussed here. Was it Haber's discovery? What were the precise roles of others? If we consider energy science and thermodynamics, the extent of those who made contributions becomes clear only in hindsight. They themselves did not always understand what they were working towards and some could not have envisioned the final outcome. We can only understand their work in historical context and through broad engagement with their research; we now consider the advances in energy science to be the work of many men over several decades (Kuhn 1959). Similarly, Nernst, in formulating the third law of thermodynamics, at first saw only the description of a set of experimental observations that served as a mathematical tool—a first application to a concrete system. Full physical insight came later, due also to the contributions of others. Determining *when* these advances were made is not always straightforward.

The case of ammonia synthesis from the elements is more clear cut, though not without ambiguity. Historically, the scientific breakthrough is seen as the work of two men, Haber and Nernst. They worked simultaneously with a clear goal of the scientific results they wished to achieve and we know exactly at which point Haber was successful on a laboratory scale and when industrial upscaling took place. Interestingly, Haber's achievement was possible without an understanding of how a catalyst functions, and without catalyst material and precursor gases whose purity would satisfy our current notion of "laboratory grade." This success illustrates a certain facet of basic scientific research: fundamental principles need not be completely understood at the outset for important results to be obtained.[2] This reality is especially important if we consider expectations of science found in the general public. The scientific endeavor is not an infallible activity of ultimate precision, it is an art form pursued by human beings, subjected to many of the same pitfalls that beset other disciplines. Guesswork and courage are absolutely necessary.

While we can learn lessons from past successes and are now more adept at "scientification" than ever before, the following holds true: "a scientific discovery must fit the times, or the time must be ripe (Kuhn 1959)." This notion seems to be a hallmark of a scientific breakthrough. It is simply expressed, but the ability to *induce* the right time and correct circumstances has so far eluded us.

We are left with an important consideration. What is the best way in today's scientific arena to move from initial identification of a phenomenon to fundamental understanding and finally to technical implementation? How do we take Hermann von Helmholtz' advice (von Helmholtz 1950),

> Whoever searches for immediate practical benefit through the study of science can be fairly certain that he will search in vain.[3]

and still later arrive at a proclamation similar to Henry Perkin after his discovery of aniline purple?

> ...the process which is now employed for [the preparation of aniline] is a remarkable instance of the manner in which abstract scientific research becomes in the course of time of the most important practical service.

It is a transition that is constantly taking place.

In today's environment of increasingly complex measurement facilities and material systems, our strategy is to bring the necessary conditions into close proximity to induce a reaction. The result is something I refer to as *The Haze*. In the case of Haber, we see how it can occur spontaneously over decades or centuries as is it thickens to a critical density out of which scientific discovery may emerge. Can we accelerate the process?

---

[2] This is often the case, however, for breakthroughs in engineering. As an example in science, see Yeang (2014, chapter 11).

[3] "Wer bei der Verfolgung der Wissenschaften nach unmittelbarem praktischem Nutzen jagt, kann ziemlich sicher sein, daß er vergebens jagen wird."

The field of chemistry, for example, can be helpful in many of the current challenges we face—either on its own or in interdisciplinary approaches with any number of other fields. These include biomedical research, climate, access to potable water, energy generation, electromobility, material recovery/separation, and recycling, or even the design of entire recycling (or other material) systems (Bender et al. 2018). Many of these will be indispensable in any kind of sustainable economy. However, the last century of research funding combined with the nature of research itself makes it difficult to know if and when research should be directed—or if it must be left completely unmanaged. This uncertainty has been exacerbated by industry's increased focus on short term gain rather than long term ventures and stability; chemical research is now often seen as a cost rather than an investment, as economic benefits are more in focus than the advancement of social interests (Whitesides 2015).

However, some of the best examples of economic and technological success in the last twenty years are based on novel work and management models which incentivize further focus on the Haze in scientific research. What is often referred to as the "Silicon Valley Model" shows the limits of (micro-) managed, large groups tackling tasks set up as well-defined problems. In many cases, a dynamic, flexible, and decentralized approach to problem solving is more conducive to fostering effective innovation. Swift reactions to the changing realities of the tech market and a corresponding development of new products have resulted in short term planning, leading to successful long term corporate strategies (Steiber and Alnge 2016, pp. 143–155). Another example, the pharmaceutical industry, exhibited enormous growth in the 1990s due to an environment of compelled innovation provided through patents of fixed-duration and the appearance thereafter of generic drugs. However, this model has not maintained its success despite sharing many of the same approaches with Silicon Valley (Franz 2017).

When applied to basic research, the Haze offers just this kind of flexibility to find solutions to problems using independent and novel approaches. The means and personnel are brought together to develop as they may in a process that has always happened in research due to the previously mentioned difficulty of directing research efficiently (nothing can replace the creativity of the proverbial discussion at the water fountain). However, we can find ways to improve efficacy. One strategy would be to rethink the approach of problem solving and whether a well-defined problem is actually needed before a solution can be found (von Hippel and von Krogh 2017). In science there is no fixed order. For catalysis research, for example, we can roughly define that for future success we must understand the exact mode of catalytic action in performance catalysts, but unforeseen discoveries in the behavior of materials and phase transitions may do more to clarify where the solution lies than anything we currently have at our disposal. Although further modifications to scientific research may be necessary so that future technical solutions (for the energy transition, for example) may be identified and implemented, we can already point to the positive effect of flexibility and the breakdown of stringent and often arbitrary rules. Of course, we must not forget the impact of serendipity and luck. The Haze must be able to respond to any contingency.

One fundamental difference between industry and research and the implementation of, say, the Silicon Valley Model, is the assessment of a product's value. Whereas in industry, healthy sales numbers can show a product's market value in real time, the value of a discovery in basic research may not become evident for decades (or ever). It is one of the largest challenges in streamlining or manipulating the Haze. Fortunately, we already have a wealth of strategies to overcome it: they are as numerous and diverse as the number of scientific and technical research groups currently in existence.

Part II
The Scientific Breakthrough

In 1903 the physical chemist Fritz Haber began work on what would become the scientific foundation of the Haber-Bosch process, that is, the industrialization of ammonia synthesis from the elements. His work was completed in 1908 with significant contributions from the physical chemist and theorist Walther Nernst. Here, in Part II, we explore in detail the events of this five-year span that took place in the scientific arena established in Part I. The narrative serves two purposes. First, a complete analysis of Haber's and Nernst's work, rooted in the fundamentals of two of the natural sciences—physics and chemistry—is missing in existing literature and provides new insight into the history of ammonia synthesis. The scientific developments are often treated only superficially and their outcome has been used as a starting point to discuss, the industrial, political, or social contexts of Haber's relationship to BASF, the transformation of the "nitrogen industry," the rise of the high pressure catalytic industry, the environmental repercussions of industrial fertilizer production, or any other number of consequences to Haber's work. These analyses are thorough and they are important. However, the physics and chemistry are rarely understood at a sufficient level, which has led to misconceptions about this episode in the history of science. The effort here is, by placing ammonia synthesis in the correct context of thermodynamics and physical chemistry, to give more complete and informative answers to several historical questions: What was Haber's motivation for taking up the problem of ammonia synthesis (Chap. 10)? What was the nature of his interaction with Walther Nernst, which had its apex at the 14th Meeting of the Bunsen Society in 1907 (Sect. 11.1.4)? And how did this interaction shape the development of science?

The second purpose of Part II is to serve as a basis to further develop the concept of *The Haze* introduced in Part I. That is, what are the factors involved in a scientific discovery? What is the dynamic, or driving force, causing particular factors to collect and form new knowledge? What separates this process in science from innovation processes in other branches such as art, politics, or business? It is a study like the one here that can illustrate the complexity of the events within a mature arena for scientific discovery as well as the relationship between the factors leading to success. Here, the relationship between experiment and theory is fundamental because it is the direct expression of the interlinking of individual sets of knowledge (epistemic exchange). Not only does the interaction between Haber and Nernst display this dynamic, but the two men's individual successes were products of their own aptitudes in the fields of physics, chemistry, and mathematics, and their ability to combine these in both theory and the laboratory.

Laying the focus on experimental and theoretical developments has meant one other deficit in the historical literature will remain unremedied: Haber's and Nernst's assistants Gabriel van Oordt, Robert Le Rossignol, Friedrich Kirchenbauer, Karl Jellinek, Friedrich Jost, and others do not receive their due recognition. It is a task I am not well equipped to tackle and will leave the work (which has already begun) to others to complete (Sheppard 2017; Szöllösi-Janze 1998a), (Sheppard 2020, p. 158). These efforts, along with the integration of Nernst's contributions into the historical narrative of ammonia synthesis, will help us break away from the allure of the "single story" in which we establish a romanticized protagonist (or antagonist): the lone scientist who devotes years to a meticulous study of a single subject in order to arrive at a scientific breakthrough (Rudwick 1985, p. 15, chapter 2, pp. 411–428), (Adichie 2009).[1] See the movie *The Fly* or *Back to the Future*, for example.

The organization of Part II is mainly chronological and is broken into sections. The first three set the stage for the scientific developments. The fourth section chronicles the major publications on ammonia synthesis from the elements as well as the 14th Meeting of the Bunsen Society in 1907. Section five recounts Haber's cooperation with BASF as it pertains to scientific developments, and section six discusses the role of physical chemistry. The final section elaborates on the general remarks on scientific developments and revisits the concept of the Haze.

One motivation remains. If such an analysis has not been done, how is it that we are aware of the historical and scientific value of Haber and Nernst's work? One answer is that the consequences of their efforts—the ways in which our lives have changed as a result—are pervasive and well-documented. Another reason is that if we turn to the testimony of any of the historical figures involved in these events (even peripherally), we are confronted with a story of deep physical insight which had a profound impact on on how science and the nature of discovery were viewed. Paul Krassa (Fig. 1), an assistant in Haber's international and colorful laboratory in Karlsruhe, described the scientific environment in which they worked (Krassa 1966). "The history of ammonia synthesis," he wrote,

> is for me not only interesting because of its eminent meaning, but rather because it—perhaps for the first time—proves the importance of cooperation for the success of a technique exhibited by the application of theory, experiments in the laboratory, and engineering technology.[2]

---

[1] It is, however, also not in general the case that a group of absolutely equal individuals contributes to a scientific breakthrough. Those involved have differing influence making "the myth of the egalitarian research collective. . .as inappropriate as the myth of the lonely genius (Rudwick 1985, p. 411)." The contributions of those involved in the discovery of ammonia synthesis were also different and unequal.

[2] "Die Geschichte der Ammoniaksynthese erscheint mir nicht nur wegen ihrer eminenten Bedeutung interessant, sondern vor allem auch deshalb, weil sie—vielleicht als erste—die Wichtigkeit der Zusammenarbeit beweist, die die Anwendung der Theorie, die Versuche im Laboratorium und die Ingenieurtechnik für das Gelingen eines Verfahrens besitzen."

Fig. 1 Paul Krassa at the Institute for Physical Chemistry and Electrochemistry at the Technical University of Karlsruhe in 1909. *Source*: Archive of the Max Planck Society, Berlin-Dahlem, Jaenicke and Krassa Collections, Picture Number VII/13.

It was not only the actors involved in ammonia research who understood their work was of historical importance; it is a sentiment echoed by other historians as well (Mittasch 1951, p. 63), (Haber 1971, p. 91), (Szöllösi-Janze 1998b, pp. 176–177), (Sheppard 2020).

Chapter 9
The State of Ammonia Synthesis at the Turn of the Twentieth Century: The Arena for Discovery

When Fritz Haber took up the problem of ammonia synthesis in 1903 there were several methods of obtaining fixed nitrogen, both naturally and through synthesis in chemical reactions. The main natural sources were Chile saltpeter (sodium nitrate from South America) and ammonium sulfate from gas works and coking (and still some guano). Although production was rising, both were limited stock supplies, not considered viable for increased future needs of the fertilizer, dye stuff or, explosives industries. Furthermore, the Chilean exports were, from the perspective of a country like Germany, precarious in that they could be cut off by a sea blockade (Schwarte 1920, pp. 537–543), (Haber 1971, pp. 84–85), (Tamaru 1991), (Stoltzenberg 1994, pp. 137–138), (Szöllösi-Janze 1998b, pp. 155–158 and citations therein), (Smil 2001, pp. 39–51), (Cushman 2013, Chapter 2).

There were two synthetic options for producing fixed nitrogen in 1903, which still contributed only small quantities to the nitrate market. They were briefly examined in Part I but are presented here in more detail to supply context for the comprehensive discussion of ammonia synthesis that follows.

The first was the electric arc in which oxygen and nitrogen were directly combined to NO or higher oxides using high temperatures from an electric discharge. That is: $N_2 + O_2 \leftrightarrow 2\,NO$ at temperatures between 2000 and 3000 °C. It was an imitation of the natural fixation process by lightning in the atmosphere (Lunge 1916, pp. 231–237, 257–269), (Mittasch 1951), (Haber 1971, pp. 85–88), (Grossmann 1974, pp. 371–372), (Holdermann 1960, pp. 48–50), (von Nagel 1991, pp. 9–12), (Stoltzenberg 1994, pp. 135–137), (Szöllösi-Janze 1998b, p. 158), (Smil 2001, pp. 53–55). The process itself had been studied since the end of the 1700s but did not advance to viable industrial application for another century. Early efficiencies were low because the arc was unstable and localized in space so only a small portion of the surrounding air was oxidized. Furthermore, the reaction is reversible and the equilibrium at high temperatures needed for appreciable oxidation rates of N to produce NO becomes unfavorable for the oxide upon cooling (as decomposition rates increase) until temperatures fall below 1000 °C. A technique was needed by which a fast reduction in temperature could preserve the oxidized nitrogen. Between

© The Author(s) 2022
B. Johnson, *Making Ammonia*, https://doi.org/10.1007/978-3-030-85532-1_9

1902 and 1903, the Norwegians Kristian Birkeland, a physics professor, and Samuel Eyde, an engineer, developed the first industrial process by using a magnetic field to disperse the electric arc into the form of a disc. Air could be blown perpendicularly through it so that not only was the NO swept away quickly from the point of high temperature, but a larger volume of nitrogen could be oxidized. Implementation of the Birkeland-Eyde process began at Notodden in 1904 and by 1906, three ovens were producing fixed nitrogen in the form of calcium nitrate (also known as Norwegian saltpeter). Before Birkeland and Eyde's collaboration, the Badische Anilin- und Soda-Fabrik (BASF) had begun work on the electric arc in 1897 led by Otto Schönherr, a chemist, and Johannes Hessberger, an electrical engineer. While they made slower progress than the Norwegians, a process was developed between 1905 and 1907 in which a meters-long arc was surrounded by tangentially spiraling air. It was also able to overcome the difficulties of rapid cooling and oxidation of large volumes of gas which had beset earlier processes. A trial plant was built in Ludwigshafen, but due to limited success, BASF replaced it in 1907 by a larger plant in Christiansand, Norway, this time in cooperation with the owners of the Birkeland-Eyde process under the name Norsk Hydro. Norway had a decisive advantage for the electricity-intensive electric arc process: hydropower was plentiful and cheap. Only under such conditions could the process be made economical. The NO created in the arc oxidized further upon cooling to form NO_2 and treatment with water resulted in the formation of nitric acid (HNO_3). This was combined with limestone to produce calcium nitrate, $Ca(NO_3)_2$, which, when concentrated to $\sim$75%, solidified and could be brought to market. It was as effective as Chile saltpeter.

In 1911, BASF pulled out of Norsk Hydro–for several years their focus had been shifting toward Haber's direct synthesis of ammonia. While the decision caused the Norwegian company considerable difficulty, it was able to prosper during World War I and into the late 1920s, at which point it converted its facilities to the Haber-Bosch process. Fritz Haber and Walter Nernst were also active in determining thermodynamic equilibria under the physical conditions required for the electric arc. It was Haber's work in this field with Adolf König that first attracted the attention of BASF in 1907 and he recounted his activities in a lecture to the German Chemical Society in 1913 (Haber 1913a), (Stoltzenberg 1994, pp. 144–151), (Szöllösi-Janze 1998b, pp. 171–175).

The second synthetic option was a multi-step approach called the (calcium) cyanamide (Kalkstickstoff) process which found its natural analogue in the bacterial fixation of nitrogen. Metals such as calcium, barium, aluminum, or manganese were reduced with nitrogen at temperatures between 700 °–1500 °C to produce fertilizer or ammonia in a process that was first investigated in the mid-1800s. Like the electric arc, an economically viable process did not emerge until the turn of the century (Frank 1903), (Schmidt 1934, pp. 177, 246–251, 337–339), (Mittasch 1951, pp. 35–38, 57–61, 87–90), (Haber 1971, pp. 88–90), (Grossmann 1974, p. 371), (Holdermann 1960, pp. 53–56, 62–65), (von Nagel 1991, pp. 13–15), (Stoltzenberg 1994, pp. 138–139), (Szöllösi-Janze 1998b, p. 159), (Smil 2001, pp. 51–52). In 1895, Adolph Frank and Nikodem Caro began experiments on calcium carbide in order to produce calcium cyanamide, a fertilizer. The availability of carbides had recently increased through production in high-heat electric arc ovens and could possibly provide an alternative production method for cyanide compounds, which were in high demand but were also patented. They were joined by Fritz Rothe in 1897 and, now on the basis of barium,[1] found that barium carbide (instead of barium carbonate) could be used to produce barium cyanamide via hydrogenation. Using this concept, Rothe developed a process in 1898 on the basis of the cheaper calcium carbide to produce calcium cyanamide via the reaction

$$CaC_2 + N_2 \rightarrow CaCN_2 + C$$

In 1899, Frank, Caro, and Rothe founded the Cyanidgesellschaft with other industrial partners, and in 1905 they began producing calcium cyanamide, a fertilizer, with the (Rothe)-Frank-Caro process at temperatures around 1100°C. Also in 1905, just before Frank and Caro, production began at the Gesellschaft für Stickstoffdünger Westeregeln at reduced temperatures using the catalytic ability of calcium chloride, discovered by Ferdinand Eduard Polzeniusz in 1901. It was, however, the Frank-Caro process that caught on. From the calcium cyanide, ammonia could be produced by steam treatment:

$$CaCN_2 + 3H_2O \rightarrow CaCO_3 + NH_3$$

Another form of the multi-step process was developed by Carl Bosch at BASF. He knew of the earlier experiments on the reduction of barium carbonate, and in 1903, he began producing barium cyanide from barium sulfate. In 1904, Alwin Mittasch came to BASF and worked closely with Bosch on the production of ammonia by treating the barium cyanide with steam. Because barium is toxic, Bosch and Mittasch investigated other nitrides and hydrides including titanium nitride, silicon nitride, titanium hydride, tantalum hydride, and uranium hydride. Experiments with aluminum nitride were also undertaken, similar to those of Ottokar Serpek, who began pilot production based on aluminum nitride with the Société Générale des Nitrures in 1909. At BASF, the use of nitrides and cyanides represented the main attempts to synthetically produce ammonia until 1907, when their interest moved to direct synthesis from the elements over the following year (Mittasch 1951, p. 90).

Like the electric arc, the cyanamide process required high temperatures so that a proximity to hydropower was advantageous (for this reason hydropower was closely linked to fertilizer manufacturing before the availability of direct ammonia synthesis). Another difficulty was the recovery of the raw materials after ammonia production in order to close the material cycle. Thanks to its relative simplicity, however, the cyanamide processes spread across North America and Europe as well as to Japan and became the main source of synthetic nitrogen on the eve of the

---

[1] The reaction $BaCO_3 + N_2 + 4C \rightarrow Ba(CN)_2 + 3CO$ was well-known and had been investigated by Margueritte and de Sourdeval in 1860

First World War (Haber 1971, p. 90). But it, too, was eventually replaced by direct synthesis from the elements.

The three processes were not completely independent. Over the course of the cyanamide process studies, Bosch and Mittasch gained valuable experience with catalysts, mixed materials, and high-pressure synthesis, which later aided them in upscaling Haber's laboratory results. It is also evident from the longer history of both the electric arc and multi-step processes that they, like direct synthesis from the elements, could only be successfully understood and implemented on an industrial scale in the context of physical chemistry (Timm 1984). Emphasizing this fact while looking back on the prior century of empirical research, Mittasch described "the thicket in which pure empiricism finally lost itself."[2] (Mittasch 1951, p. 57). It was knowledge from physical chemistry that revealed the way out.

When Haber took up ammonia synthesis in 1903, there had already been attempts at direct synthesis, including some based on physicochemical principles (Mittasch 1951, pp. 44–45, 48–53), (Sheppard 2020, pp. 87–91).[3] In 1876, Marcelin Berthelot reported finding the same amount of ammonia in both generation and dissociation experiments using an electric spark. It was the observation of what we today call equilibrium between ammonia, hydrogen, and nitrogen. However, the discussion of whether an equilibrium actually existed would last another thirty years. A more advanced step in this direction came in 1884, when William Ramsay and Sydney Young observed the reversibility of the reaction: they showed that ammonia never completely disappeared in dissociation experiments at high temperatures. They investigated the effect of gas flow-rates, different catalytic materials and surface areas, and also attempted experiments under pressure (Ramsay and Young 1884). In 1901, Henry Le Chatelier, an early advocate of physical chemistry, investigated the synthesis of ammonia under pressures up to 100 bar at 600°C using an iron catalyst (Chatelier 1888), (Chatelier 1936, pp. 73–76). His detonation experiments with a mixture of hydrogen and nitrogen (he ended up destroying the experimental apparatus in the process) were used to obtain a patent in 1903, but his positive results were erroneous. Nevertheless, his work is notable for its reliance on physicochemical principles [compare: Silverman (1838), Tamaru (1991), Mittasch (1951, pp. 52, 54)]. Later, Le Chatelier would, like Mittasch, distinguish strictly between the new chemistry and the approach that had preceded it.

Wilhelm Ostwald and his assistants also took up a serious investigation of ammonia synthesis from the elements in 1900 at his institute in Leipzig (Figs. 9.1 and 9.2). They used a 1-to-3 nitrogen-to-hydrogen ratio, under pressure, with temperatures between 250 and 300°C, and mainly iron and copper catalysts with large surface areas (Ostwald 1927, pp. 279–287), (Farbwerke Hoechst 1964), (Holdermann 1960, pp. 40–42). After they believed to have synthesized ammonia— according to Ostwald he achieved the impressive yield of 8% although lesser

---

[2] ". . .in welches Dickicht sich die reine Empirie schließlich verlor. . ."

[3] For a more comprehensive overview of ammonia synthesis experiments in the nineteenth century see (Mittasch 1939) and (Mittasch 1951).

amounts had also been observed (Ostwald 1927, p. 286)—he filed for a patent in March and entered into negotiations with the Badische Anilin & Soda Fabrik (BASF), the Farbwerke Hoechst in Frankfurt, and the Elberfelder Farbwerke to industrially upscale the process. Ostwald received several lucrative offers and test production began. In the ensuing investigations, Carl Bosch, who had come to BASF in 1899, found the iron in Ostwald's process was not acting as a catalyst, but instead contained atomic, reactive nitrogen left over from production (Farbwerke Hoechst 1964, pp. 22–25). As a test, Bosch placed two different iron wires in glass pipes and passed pure hydrogen over one, and a 1:3 N_2:H_2 mixture over the other between 350 °C and 450 °C. The vessel containing pure hydrogen yielded 6.2 mg ammonia while the vessel containing the N_2/H_2 mixture yielded less, only 3.6 mg. In other words, the configuration that ideally should have produced no ammonia yielded more than the configuration of the process Ostwald purported to have developed. The ammonia had been generated when hydrogen passed over the heated iron already containing nitrogen. According to literature, Bosch reported, red-hot iron absorbed between 0.2% and 11.5% nitrogen when ammonia was dissociated over it or when nitrogen was passed over it. When cooled in the presence of hydrogen, the absorbed nitrogen reacted slowly to ammonia. Bosch requested Ostwald's original experimental apparatus and tested it directly. He heated it to 400°C, passed pure hydrogen over the iron catalyst for 25 hours and achieved "relatively high amounts of ammonia."[4] This result sealed the deal for Bosch and similar conclusions were soon reported by Hoechst. It was another in a long line of mistaken detections of ammonia synthesized from the elements (Mittasch 1951).

While Bosch sought clarification from Ostwald to help with the continuation of the experiments, Ostwald decided not to continue in the field of ammonia synthesis. In light of Bosch's work, he withdrew his patent application and backed out of the agreements with industry. He turned to the synthesis of nitric acid from ammonia, hastened by a personal belief that if cut off from Chile saltpeter, Germany could find itself in a precarious situation on the world stage (Ostwald 1903), (Ostwald 1927, pp. 287–288).

Despite the negative result of his experiments, we consider here several passages from his patent application. "I have found," Ostwald wrote (Ostwald 1927, pp. 284–285),

> that the binding of free nitrogen with hydrogen through a suitable contact substance or catalyst at low temperatures of 250 ° to 300 ° can be realized at measurable rates. The rate increases quickly with increasing temperature. Metals for example, mainly iron and copper, having large surface areas, can be used as catalysts. The binding is never complete, but rather leads to a chemical equilibrium and the amount of ammonia formed is, therefore, dependent on the ratios of the precursors. In order to achieve complete conversion, one must remove the ammonia from the reaction mixture, which can be realized through dissolving

---

[4] "...immer noch eine verhältnismässig reichliche Menge Ammoniak."

Fig. 9.1 Wilhelm Ostwald in 1902. *Source*: Archiv der Berlin-Brandenburgischen Akademie der Wissenschaften, NL W. Ostwald, Nr. 5292

Fig. 9.2 Portrait of Ostwald in 1898 at the time Haber was inquiring about a post in the senior scientist's institute. By David Vandermeulen from *Fritz Haber: L'Esprit du Temps* (Vandermeulen 2005); © David Vandermeulen/Guy Delcourt Productions

the ammonia in water or acid. For this, the mixture can be circulated, if necessary while cooling it to re-use the heat contained therein.[5]

---

[5] "Ich habe gefunden, daß die Verbindung von freiem Stickstoff und Wasserstoff durch geeignete Kontakt- substanzen oder Katalysatoren bereits bei geringer er-hitzung auf 250° bis 300° mit meßbarer Geschwindigkeit bewirkt werden kann. Die Geschwindigkeit nimmt mit steigender Temperatur schnell zu. Als Katalysatoren dienen beispielsweise Metalle, hauptsächlich Eisen und Kupfer, denen man große Oberflächen gibt. Die Ver-bindung ist nie vollständig, sondern führt zu einem chemischen Gleichgewicht und die gebildete Ammoniak-menge ist daher von dem Mengenverhältnis der Stoffe abhängig. Um die Verbindung vollständig zu machen, muß man das Ammoniak aus dem Reaktionsgemisch ent-fernen, was durch Aufnahme desselben mit Wasser

Later he continued: "Because the relative amount of ammonia in the gas mixture increases with increasing pressure, it is advisable to conduct the synthesis under high pressure."[6]

With these statements Ostwald succinctly wrote the "recipe" for ammonia synthesis from the elements and as a result, identified himself as the "intellectual father of [the nitrogen] industry (Ostwald 1927, p. 285)."[7] While the recipe contained the same concepts used later by Haber and Nernst in their successful experiments, a look at the breadth of attempts to synthesize ammonia during the nineteenth century shows that all of these factors, if only separately, had already been discussed at that time in the literature in terms of physicochemical principles. One possible exception was a system of heat exchange (Mittasch 1951). This illustrates again that a holistic (or global) approach was still missing in attempts to synthesize ammonia before 1900. Although Ostwald brought these elements together, he lacked both a quantitative theoretical underpinning of chemical equilibria, later provided by Nernst, and a sensitivity to the reproducibility of small quantities of ammonia in equilibrium with hydrogen and nitrogen at normal (or even higher) pressures, an achievement of Haber's. Several pages from Ostwald's laboratory book show values from a review of ammonia experiments obtained in pressure oven experiments from March and April of 1900 while he was in negotiations with BASF and Hoechst (Figs. 9.3 and 9.4) (Ostwald 1900, pp. 23–25).[8] The equilibrium values for ammonia fluctuate widely compared, for example, to Ramsay and Young (although they, too, complained of their own "want of uniformity"). The latter study, on the other hand, reported values for the percentage of decomposed ammonia at raised temperatures which are too high—at 830 °C/1100K and 1 atm well over 99% should be decomposed (see Ramsay and Young (1884) and Fig. 9.5). As is evident from his publications, it was Haber who was first able obtain the accuracy and reproducibility needed for reliable measurements.

There was one other modern effort to synthesize ammonia from the elements at the turn of the twentieth century: that of Edgar Perman. His work will be discussed with Haber's first publications in the context of their exchange in literature.

We know today that none of these attempts to synthesize ammonia from the elements could come to fruition and no other sources of fixed nitrogen, neither natural nor synthetic, would prove as viable as the Haber-Bosch process at that

---

oder Säuren geschehen kann. Das Gasgemisch kann zu diesem Zweck einen Kreislauf, nötigenfalls unter Abkühlung und Wiedergewinnung der Wärme durchmachen."

[6] "Da die verhältnismäßige Menge des Ammoniaks im Gasgemische mit steigendem Druck zunimmt, so ist es zweckmäßig, die Synthese unter vermehrtem Druck aus-zuführen."

[7] "...der geistige Vater [der Stickstoff-] Industrie..."

[8] A letter of recollection from Ostwald's assistant Eberhard Brauer is also contained in the Archiv der Berlin-Brandenburgischen Akademie der Wissenschaften (Brauer 1954). He shared his memories on the attempts at synthesizing ammonia in 1954 because the documents from the research on ammonia and nitric acid were lost in a cellar flood in 1938.

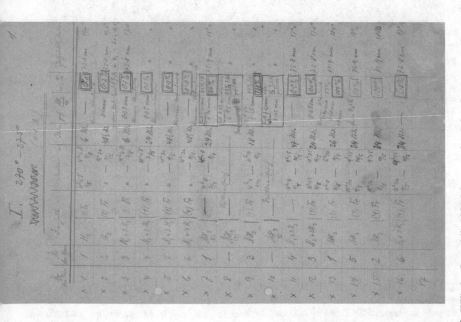

Fig. 9.3 Wilhelm Ostwald's 1900 laboratory book for nitrogen experiments in a pressure oven, pages 23 and 24. *Source*: Archiv der Berlin-Brandenburgischen Akademie der Wissenschaften, NL W. Ostwald, Nr. 4397–3

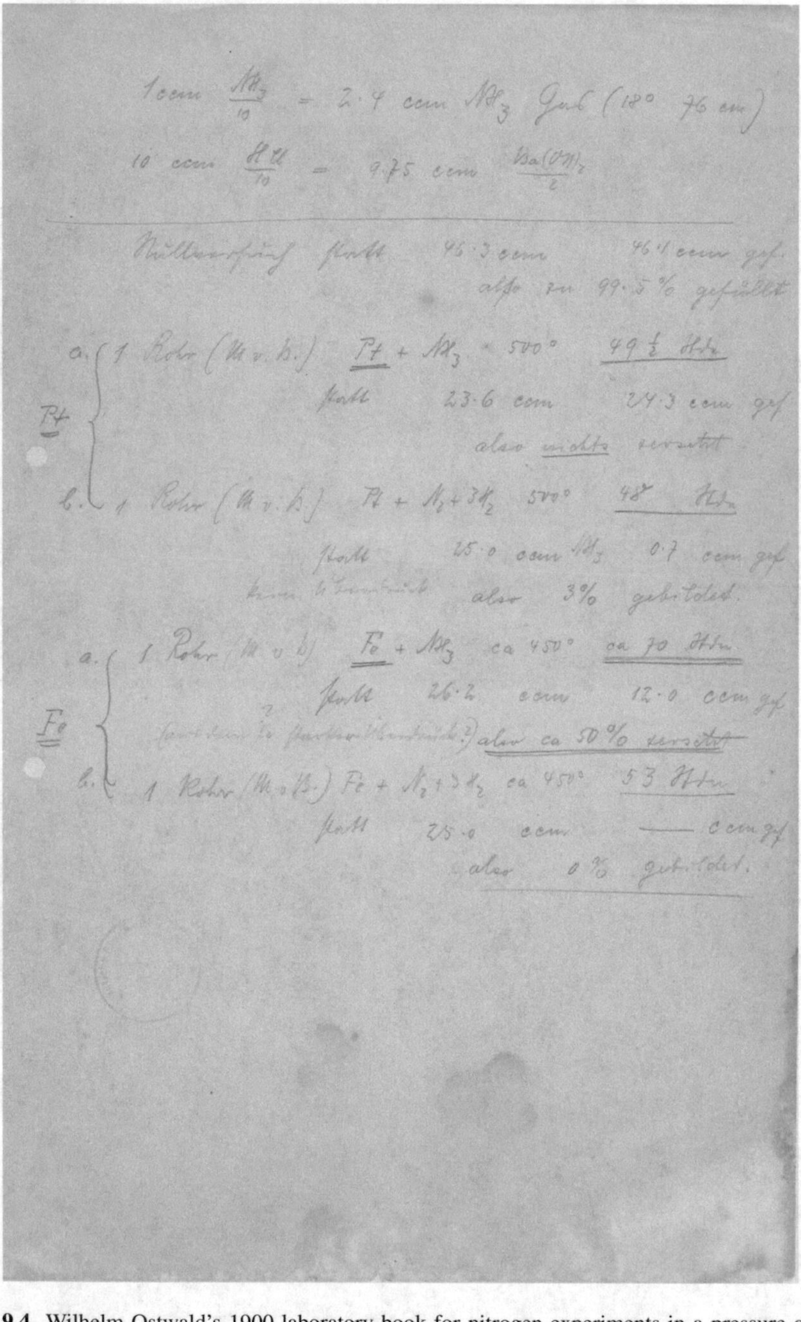

Fig. 9.4 Wilhelm Ostwald's 1900 laboratory book for nitrogen experiments in a pressure oven, page 25. In the experiments with a platinum catalyst marked "Pt a." and "Pt b." the conditions are nearly the same: 500 °C at $49\frac{1}{2}$ Atm and 48 Atm, respectively. In the former, no ammonia was dissociated; in the latter, 3% was generated from a starting mixture of N_2 and H_2. Theoretically, the values should have been nearly the same. In the iron experiments, "Fe a." and "Fe b.", at 450 °C, 50% of the ammonia was dissociated at 70 Atm while at 53 Atm no ammonia was formed from the N_2+H_2 mixture. The latter result illustrates the difficulty of combining nitrogen and hydrogen. *Source*: Archiv der Berlin-Brandenburgischen Akademie der Wissenschaften, NL W. Ostwald, Nr. 4397–3

90   RAMSAY AND YOUNG ON THE

Temperature.	Weight of ammonia decomposed.	Weight of ammonia undecomposed.	Percentage of ammonia decomposed.	Mean.
600°	0·04493	0·19880	18·43	18·28
,, ...	0·04486	0·20250	18·14	
620	0·04182	0·13446	23·72	
,, ...	0·04503	0·11502	28·79	25·58
,, ...	0·04560	0·14255	24·24	
680	0·04517	0·08586	34·47	
,, ...	0·04636	0·08910	34·22	35·01
,, ...	0·04629	0·08100	36·36	
690	0·04542	0·05832	43·78	
,, ...	0·04542	0·04212	51·88	47·71
,, ...	0·04542	0·05022	47·49	
810 to 830° ..	0·04431	0·02754	61·67	
,, ,, ..	0·04772	0·01620	74·66	
,, ,, ..	0·04613	0·01134	80·27	
,, ,, ..	0·34285	0·14256	70·63	69·5
,, ,, ..	0·34561	0·20412	62·87	
,, ,, ..	0·36264	0·15390	67·04	

The want of uniformity in the results at the higher temperatures appears to be due to two causes—(1) the difficulty in keeping the temperature of the tube constant; thus in the last series the tube was at one time very nearly at 810°, and at another at about 830°; (2) The variable rapidity of the current of gas, which was found subsequently to exercise great influence on the amount of decomposition, and this might be anticipated, since there is practically no recombination of nitrogen and hydrogen to form ammonia.

A second series of experiments was made with an iron tube filled with broken pieces of porcelain.

Temperature.	Weight of NH_3 decomposed.	Weight of NH_3 undecomposed.	Percentage of ammonia decomposed.	Mean.
507—527°	0·04432	1·01410	4·19	4·15
,, ...	0·02303	0·53622	4·12	
600° (a)........	current very fast.			
,, ...	0·04644	0·17658	20·82	
,, ...	0·04540	0·17344	20·76	21·36
,, ...	0·04527	0·15522	22·49	
,, (b)........	current much slower.			
,, ...	0·04527	0·08586	34·44	34·44
628	0·04733	0·02268	67·61	
,, ...	0·04558	0·02592	63·75	65·43
,, ...	0·04558	0·02106	68·40	
,, ...	0·34980	0·21465	61·97	

Fig. 9.5 Two pages from Ramsay and Young's 1884 publication "The Decomposition of Ammonia by Heat" showing their results using iron filings (Ramsay and Young 1884). The consistency of the measurements is evident. *Reproduced by permission of The Royal Society of Chemistry*

DECOMPOSITION OF AMMONIA BY HEAT.   91

Temperature.	Weight of NH₃ decomposed.	Weight of NH₃ undecomposed.	Percentage of ammonia decomposed.	Mean.
676—695°	0·35407	0·16524	68·02	
,, ...	0·36473	0·19926	64·67	66·57
,, ...	0·35762	0·17334	67·35	
,, ...	0·35941	0·18306	66·25	
730	0·34840	0·02268	93·89	
,, ...	0·35459	0·02106	94·39	93·38
,, ...	0·35034	0·02268	93·92	
,, ...	0·35850	0·03402	91·33	
780	vol. = 1 litre	—	100	100
,, ...	,,	—	100	

It is to be remarked that in all these experiments with broken porcelain, the surface became blackened, owing, doubtless, to the reduction of the lead used in the enamelling. The surface was consequently blistered, and possibly to some extent porous.

The gas was then passed through a plain glass combustion tube lying in the iron tube, which was heated to 780°.

Weight of ammonia decomposed.	Weight of ammonia undecomposed.	Percentage of ammonia decomposed.
0·00571	2·4056	0·24

That the small quantity of gas collected was really a mixture of nitrogen and hydrogen was proved by its burning quietly with a colourless flame as usual.

The ammonia was then passed through a glass tube quite filled with fragments of broken combustion tubing to increase the surface. The temperature was again 780°, but rose slowly.

Weight of ammonia decomposed.	Weight of ammonia undecomposed.	Percentage of ammonia decomposed.	Mean.
0·01016	0·7776	1·29	
0·01020	0·6042	1·66	1·72
0·01040	0·5540	1·84	
0·01720	0·8019	2·10	

A glass tube was then filled with strips of ignited asbestos cardboard, which being porous should expose a large surface to the gas.

Temperature.	Weight of ammonia decomposed.	Weight of ammonia undecomposed.	Percentage of ammonia decomposed.	Mean.
520°	0·00788	0·2641	2·90	2·90
780	vol. = 1 litre	—	100	100
,, ...	,,	—	100	

H 2

Fig. 9.5 (continued)

time.[9] Some (estimated) numbers by L.F. Haber put the world's sources of fixed nitrogen into perspective in both 1900 and in 1913 after synthetic production emerged on the market (Haber 1971, pp. 101–104).[10] In 1900, fixed nitrogen production from coking (ammonium sulfate) was 85,000 metric tons while 230,000 tons came from nitrate deposits. In 1913, out of a total consumption of 750,000 metric tons of fixed nitrogen, 280,000 tons (37%) were from coking, 430,000 (57%) tons from nitrate deposits, 24,000 tons (3%) from the cyanamide process, and 15,000 tons (2%) from the electric arc. The first Haber-Bosch plant at Oppau began production in late 1913 and by 1914 was producing ammonium sulfate at the rate of 36,000 tons per year, which amounted to about 8,000 tons of fixed nitrogen. By 1917/1918, 105,000 tons of nitrogen were being produced annually in Germany using the Haber-Bosch process, representing about 40% of total production; this number rose to three quarters over the next decade. In 1930, the German chemical industry was producing 1.2 millions tons (Mittasch 1951, pp. 136–137), (Szöllösi-Janze 2000, pp. 119–121). In comparison, at the end of World War I, saltpeter exports were also in the millions of tons (Eucken 1921, p. 170). In 2005, total global man-made nitrogen production was 140 million tons, of which 85 million (~60%) was manufactured with the Haber-Bosch process (Erisman et al. 2005).

---

[9] With current technology, other options for producing synthetic ammonia are being developed. See, for example, Power-to-X technology (Ausfelder and Durra 2019; Service 2018).

[10] W. Eucken (1921)[pp. 28–54] gave similar numbers and trends.

Chapter 10
Fritz Haber's Work and Thought as He Began Work on Ammonia Synthesis

Fritz Haber was born in Breslau (today Wrocław, Poland) in 1868 (Coates 1937), (Stoltzenberg 1994, chapters 2, 3, 4), (Szöllösi-Janze 1998b, chapters 1, 3), (Sheppard 2020, Chapter 2) (Figs. 10.1 and 10.2). He grew up in a prominent Jewish family involved in the chemical and dye trade, and received his education at one of the city's normal *Gymnasia*. Near the end of his schooling, he became interested in mathematics and the physical sciences, and although chemistry was still only a small part of the curriculum, it was this subject that caught Haber's attention, and he embarked on a path in higher education that "deviated strongly from the normal pattern (Schlenk 1934)."[1] In 1886, he completed his *Abitur* and began studying chemistry, first at the Friedrich Wilhelm University in Berlin, then at the University of Heidelberg, and then again in Berlin at the Technical University of Charlottenburg; in 1889, he began his one year military service in Breslau. The plan thereafter was to begin studying organic chemistry at the Friedrich Wilhelm University as preparation for a career in the coal tar dye industry. However, as Haber resumed his studies, he was not convinced of his choice. He conducted research in organic chemistry and received his Ph.D. in 1891, but his future was still uncertain. Haber spent time in industry and then in academics before joining his father's dye and chemical business. Only after that did he make the decisive turn back toward an academic career and begin working in organic chemistry at the University of Jena in 1892. It was also there that he converted to Christianity. However, Haber remained unsatisfied by the challenges of his ordinary laboratory tasks; it was his introduction to the emerging discipline of physical chemistry that intrigued him. Despite his inquiry, Haber was not invited to join Ostwald's institute in Leipzig so he instead went to the Technical University of Karlsruhe in 1894 to work with Hans Bunte and Carl Engler, who was closely involved with the coal tar dye industry. He remained there for 17 years—it was a decision that would have lasting consequences.

---

[1] "...vom üblichen Schema stark abwich."

© The Author(s) 2022
B. Johnson, *Making Ammonia*, https://doi.org/10.1007/978-3-030-85532-1_10

Fig. 10.1 Breslau in 1891, the home town of Fritz Haber. By David Vandermeulen from *Fritz Haber: L'Esprit du Temps* (Vandermeulen 2005); © David Vandermeulen/Guy Delcourt Productions

As assistant to Bunte in the department of Chemical Fuel and Technology, Haber worked mainly on organic chemical problems, and in 1896 he published his *Habilitationsschrift* on the thermal dissociation of hydrocarbons (Haber 1896). As *Privatdozent*, and with the help of his friend Hans Luggin, Haber began work on electrochemical systems and finally entered the field of physical chemistry in both experiment and theory. In 1898, he published a book on the subject and in

Fig. 10.2 Prof. Fritz Haber in 1909, probably in a lecture hall at the Technical University of Karlsruhe. *Source*: Archive of the Max Planck Society, Berlin-Dahlem, Jaenicke Collection, Picture Number I/4

the same year became professor extraordinarius (Haber 1898). Three years later, Haber married Clara Immerwahr, who had written her dissertation in chemistry under Richard Abegg in 1900; she was the first woman at the University of Breslau to receive her Ph.D (von Leitner 1993, pp. 51–68), (Friedrich and Hoffmann 2017b) (Figs. 10.3 and 10.4).

Why, at this time in his career, did Haber decide to take up the problem of ammonia synthesis? He was a unique chemist of varied interests, who overcame the divide between organic and physical chemistry, in part due to his ability to understand the required mathematics (Haber 1905b),[2] (Engler et al. 1909), (Ostwald 1927, pp. 111–113), (Haberditzl 1960; Planck 1946), (Szöllösi-Janze 1998b, p. 94). This would later provide the key to his success, but was also important to his motivation.

---

[2] Haber's 1905 book *Thermodynamics of Technical Gas Reactions* was published in English in 1908 (Haber 1908d).

Fig. 10.3 Clara Haber, née Immerwahr, undated. *Source*: Archive of the Max Planck Society, Berlin-Dahlem, Jaenicke and Krassa Collections, Picture Number V/3

Fig. 10.4 Portrait of Clara Haber, née Immerwahr by David Vandermeulen from *Fritz Haber: L'Esprit du Temps* (Vandermeulen 2005); © David Vandermeulen/Guy Delcourt Productions

Authors of both academic and popular literature have remarked that Haber chose to work on the equilibrium of ammonia for financial gain (as an advisor to industrial partners) or that he viewed the N_2-H_2-NH_3 system as a fundamental physical problem. While these factors are correct, they are often of vague origin and Haber's actions are imbedded in a turn-of-the-century attitude about the need for a synthetic source of nitrogen (Coates 1937, pp. 1650–1651), (Mittasch 1951, p. 64), (Wendel 1962, pp. 23–27), (Goran 1967, pp. 42–45), (Szöllösi-Janze 1998b, p. 159), (Charles 2005, pp. ix-vx, 81–85), (Erisman et al. 2008; Hager 2008). Fixed nitrogen could be used to manufacture explosives and a synthetic source would be crucial if sea routes transporting nitrates from South America to Europe were disrupted by hostile powers (Ostwald 1903; Tamaru 1991). It was also necessary for fertilizer production to guard humanity against impending famine, a sentiment embodied by William Crookes' 1898 appeal in which he also called attention to the economic potential of a synthetic nitrogen source (Crookes 1917, pp. 2–3, 37–38). What part did this goal play in Haber's motivation?

An example of this ambiguity is found in Haber's own 1920 Nobel lecture (Haber 1920, pp. 1–2):

> A narrow professional interest in the preparation of ammonia from the elements was based on the achievement of a simple result by means of special equipment. A more widespread interest was due to the fact that the synthesis of ammonia from its elements, if carried out

on a large scale, would be a useful, at present perhaps the most useful, way of satisfying important national economic needs. Such practical uses were not the principal purpose of my investigations. I was never in doubt that my laboratory work would produce no more than a scientific confirmation of basic principles and a criterion of experimental aids, and that much would need to be added to any success of mine to ensure economic success on an industrial scale. On the other hand I would hardly have concentrated so much on this problem had I not been convinced of the economic necessity of chemical progress in this field, and had I not shared to the full Fichte's conviction that while the immediate object of science lies in its own development, its ultimate aim must be bound up in the moulding influence which it exerts at the right time upon life in general and the whole human arrangement of things around us.

Since the middle of the last century it has become known that a supply of nitrogen is a basic necessity for the development of food crops; it was also recognized, however, that plants cannot absorb the elementary nitrogen which is the main constituent of the atmosphere, but need the nitrogen to be combined with oxygen in the form of nitrate in order to be able to assimilate it. This combination with oxygen can start with combination with hydrogen to form ammonia since ammonium nitrogen changes to saltpetre nitrogen in the soil.

While Haber mentioned several reasons why he took up ammonia synthesis, they are obscured by the context of a need for a synthetic source of fixed nitrogen as his speech continues. Perhaps Haber simply did not recall his initial motivation– this situation would not be unique (Fleck 1980, pp. 95, 100–102, 111), (Rudwick 1985, p. 7), (Holmes 1991, pp. 307–343), (Blum et al. 2017; Graßhoff 1994, 1998). However, by considering Haber's work and correspondence at the turn of the twentieth century, we can establish a finer understanding of how he came to take on the task.

During his early years in Karlsruhe, Haber had varied organic and physico-chemical interests "of a downright baffling variety"[3] and yet he was still able to investigate the subjects in great detail (Jaenicke 1958b; Schlenk 1934), (Sheppard 2020, pp. 53–54). They included combustion, electrolysis, heats of reactions, and pulverization. He also maintained close ties to industry. These interests remained when he took up ammonia synthesis in 1904, as is evident in the publications which appeared soon after. Some had direct bearing on ammonia synthesis, for example his work on the free energies of chemical reactions (Haber 1905b, Chapters 1–3) and electric arcs (Haber and Moser 1905c; Haber and Bruner 1904g; Haber and Richardt 1904h), while electrochemistry (Haber and Schwenke 1904i; Haber and Tołłoczko 1904j), the state of technological innovation, and education were also high priorities (Haber 1902b,c, 1903a,b,d). Haber, the trained experimentalist, had a life-long ambition to move toward ever more abstract and theoretical studies and away from his experimental roots. "In my early years," wrote the autodidact Haber retrospectively in 1911 (Haber 1911),

---

[3] "Von geradezu rätselhafter Vielseitigkeit..."

I had no understanding of physical chemistry or physics and so I was forced to learn these things on the side during other work in later years as I moved from engineering to technology and from there to theory...[4]

This attitude can also be seen in Haber's relationship with Einstein (Renn 2006, pp. 73–78). It was a development that began during the time physical chemistry was emerging as a distinct discipline: the strong connections to organic chemistry that appear in Haber's early work gave way to more esoteric topics as he moved toward the new field (Jaenicke 1958a), (Coates 1937, p. 1643), (Girnus 1987).

In 1904, for example, Haber published "Über die kleinen Konzentrationen" (On small concentrations) (Haber 1904f) (see also Haber (1904e)), in which he attempted to reconcile experimental observations with physical laws. It is an important illustration not only of Haber's progress as a physical chemist (the problem occupied many of the field's top researchers) but also of how far some of his interests lay from ammonia synthesis. The paper was based on the observation that the theories of Jacobus Henricus van't Hoff, Svante Arrhenius, and Wilhelm Ostwald for applying thermodynamics to solutions—the basis of physical chemistry (Ertl 2015)—became more precise the more the solution tended toward "infinite" dilution. In the limiting case, the dissociation of the molecules in the solution was complete and fully described by theory (Arrhenius 1887; van't Hoff 1887), (Windisch 1892, pp. 476–522), (Ostwald 1927, pp. 16–31), (Partington 1964, pp. 637–662, 663–681). For Haber, however, measurements of the densities of ions in these dilute solutions posed a serious problem with respect to the speed of light. In a well-known system of dilute silver salts, the concentration of the silver ions, determined from voltage measurements, was so small Ostwald came to the conclusion that only two free silvers atoms existed in every cubic centimeter of solution. If one took such a cubic centimeter and divided it into three parts, at least one would contain no silver ions at all—but only if the ion remained dissociated forever. According to Ostwald, such an assumption would be "completely inappropriate." He continued (Ostwald 1893),

Rather, we must interpret the result in such a way, that in general every silver atom exists only temporarily in an ionic state and that its lifetime in that state is only [a] fraction...of the lifetime in the state of the [silver] complex.[5]

According to Haber, this explanation was the only way to present the results in an atomistic way. The problem emerged when he examined the ratio of the decay time to the generation time of the silver complex in equilibrium using a model based on the kinetics in Ostwald's assumption. The answer was that the complex was

---

[4] "...in meinen jungen Jahren [habe] ich weder von der physikalischen Chemie noch von der Physik eine Kenntnis gehabt und so habe ich, indem ich mich von der Technik zur Technologie und von dieser zur Theorie gewandt habe, immer neben anderer Arbeit die Dinge in reiferen Jahren zulernen müssen...".

[5] "...ganz unangemessen. Vielmehr müssen wir das Ergebnis so auffassen, dass im allgemeinen jedes Silberatom vorübergehend in den Ionenzustand gelangt, dass aber seine Existenzdauer in diesem Zustande nur [ein] Bruchteil...von seiner Existenzdauer im Zustand des Komplexes...ist."

generated 9×10^{21} times faster than it decayed. Consideration of several realistic decay times led Haber to the conclusion that the generation time was on the order of 10^{-24} s. However, this made no more sense than having only two silver ions per cubic centimeter of solution. The formation of a complex, Haber argued, was due to the rearrangement of electric charge and required the movement of matter. But light propagated across atomic dimensions in 10^{-18} s—a million times slower than the formation of the silver complex. And so, Haber wrote,

> it is not apparent how such a change should be able to take place more quickly than electrical changes otherwise proceed...[which is] at the speed of light...[6]

The measurements of the ion densities in other dilute solutions were even more egregious. Haber did not propose a solution, except that the measured potentials used to determine the ion concentrations were perhaps only operands (Rechengrößen) and did not correspond to actual concentrations in solution.

By 1904, the topic of dilute solutions had become a discussion in the literature between Haber, Guido Bodländer, Heinrich Danneel, and Richard Abegg (Abegg 1904b; Bodländer 1904a; Bodländer and Eberlein 1904b; Danneel 1904). "It is not my intention," Haber wrote to Ostwald on the topic, "to assert anything for or against the atomistic or energetic view with these considerations."[7] Later the same year, he expressed further thoughts in continuing communication with Abegg (Haber 1904b). The pertinence of the discussion of the atomistic-versus-energetic-view is notable considering Einstein published his theoretical treatment of Brownian motion the next year and Jean Perrin's experimental confirmation came in 1909 (Einstein 1905; Perrin 1909).

A similar example of Haber's extensive interest in reaction rates and dilute solutions can be found in another 1904 paper, "On the Electrochemical Determination of the Susceptibility of Glass," as well as in correspondence with Ostwald (Haber 1904a,d; Haber and Schwenke 1904i), and in ongoing discussions with Abegg on the latter's related theory of valency (Abegg and Bodländer 1899; Abegg 1904a; Haber 1906). He must have been in the middle of these considerations when he received a letter from the Margulies brothers of the Österreichische Chemische Werke in Vienna requesting his advice on the catalytic synthesis of ammonia (apparently they also visited Haber personally in Karlsruhe (Jaenicke 1958b)). As is well known, this communication started him on the path to ammonium synthesis from the elements. In 1903, Haber wrote to Wilhelm Ostwald about the issues raised by the Margulies, asking whether the senior scientist would like to take on the task himself (Haber 1903c) (Fig. 10.5). It is informative to quote the entire letter.

> Hoch geehrter Herr Geheimrat [Ostwald]!
>   An Austrian company has recently written to me repeatedly to hear my opinion on whether it would be worth their while to catalytically produce ammonia from nitrogen

---

[6] "...es ist nicht zu erkennen, wie eine solche Veränderung rascher sollte geschehen können als elektrische Veränderungen sonst irgendwie vor sich gehen...mit der Lichtgeschwindigkeit...".

[7] "Es ist nicht meine Absicht mit diesen Betrachtungen etwas für oder wider die atomistische oder die energetische Betrachtungsweise vorzubringen...".

and hydrogen on a large scale. I answered by calling their attention to the low costs of producing ammonia as a byproduct of coking plants. This advice did not decrease the company's inclination toward producing ammonia via the above mentioned process. A foreign colleague approached me at a conference in Berlin and said that you, hoch geehrter Herr Geheimrat, have studied this question and have worked out a technically useful method. Therefore, please let me know whether it would suit you if I make the necessary arrangements for the Austrian company to contact you. I am also driven by the thought that you may find it desirable that a financially strong and reputable company take over the technical implementation. At the same time, nothing could be as valuable for this company as to have you as a consultant. I would like to add, however, that I, in the event that you do choose to take up this venture, cannot vouch for the technical implementation. Neither is my relationship to the Austrian company close enough to have any influence, nor do I wish to hide the fact that industrial undertakings inside Austrian territory are often planned but seldom realized–even by the most reputable and powerful companies.[8]

Mit vorzüglicher Hochachtung bin ich Ihr Ergebener

Haber

While Haber must have been considered knowledgeable enough on the matter of catalytic ammonia synthesis for the company to contact him, he was not familiar with the current state of research in 1903. If Ostwald had worked out a usable solution, Haber might have expected to hear about it from Ostwald himself, or, if he had been convinced of the magnitude of such a discovery, to have read about it in the literature. Either way, he did not seem particularly interested in the undertaking. This is not to say he had no understanding of the role of fixed nitrogen in fertilizer and explosives, but in 1903 his interest lay foremost in the industrial side of the operation. For Haber, a technical solution depended on the interplay between market forces and the constraints placed on the system by nature–knowledge recently made accessible through physicochemical concepts. He did not see it as a technical development that must be desperately pursued and hastily implemented. Furthermore, as would remain the case for several years, Haber considered ammonium sulfate from coking plants and the electric arc to be more viable options of producing fixed nitrogen than direct synthesis from the

---

[8] "Eine Österreichische Firme hat mir wiederholt in neuerer Zeit geschrieben, um meine Meinung darüber zu hören, ob es sich für sie empfiehlt Stickstoff und Wasserstoff im großen katalytisch zu Ammoniak zu vereinigen. Darauf habe ich mit einem Hinweis auf die niedrigen Kosten der Ammoniakgewinnung im Nebenprodukt [der] Kokerei-Betriebe geantwortet. Dieser Hinweis hat die Neigung der Firma nicht vermindert die Ammoniakbereitung auf dem erwähnten Weg zu verfolgen. Nun hat mir ein ausländischer Kollege beim Kongress in Berlin erzählt, daß Sie, hoch geehrter Herr Geheimrat, diese Frage studiert und bis zur Ausarbeitung eines technisch gangbaren Weges gefördert hätten. Ich bitte Sie deshalb mich wissen zu lassen, ob es Ihnen angenehm ist, wenn ich die Österreichische Firma veranlasse, sich an Sie zu wenden. Es leitet auch dabei der Gedanke, daß es Ihnen vielleicht erwünscht ist, wenn eine kapitalkräftige und ehrenwerte Firma die technische Durchführung in die Hand nimmt, während dieser Firma nichts wertvoller sein könnte, als Sie zum Berater zu haben. Ich möchte aber hinzufügen, daß ich, gesetzt den Fall, daß Sie auf die Sache eintreten, für die schließliche Durchführung nicht gut sagen kann, da ich weder dem österreichischen Hause nahe genug stehe, um einen entscheidenden Einfluß zu üben noch verhehlen möchte, daß industrielle Unternehmungen auf österreichischem Boden auch von den solidesten und stärksten Firmen oft geplant und ziemlich selten realisiert werden."

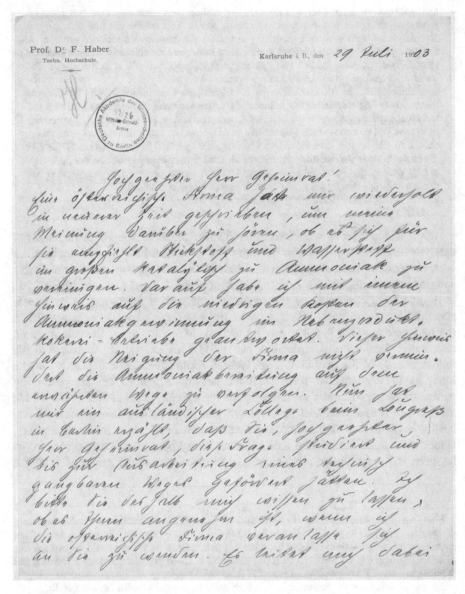

Fig. 10.5 The letter Fritz Haber sent to Wilhelm Ostwald on June 29, 1903 in which Haber told of the Margulies brothers' interest in industrial catalytic ammonia production and asked whether Ostwald would be interested in a cooperation with them (Haber 1903c). *Source*: Archiv der Berlin-Brandenburgischen Akademie der Wissenschaften, NL W. Ostwald, Nr. 1037

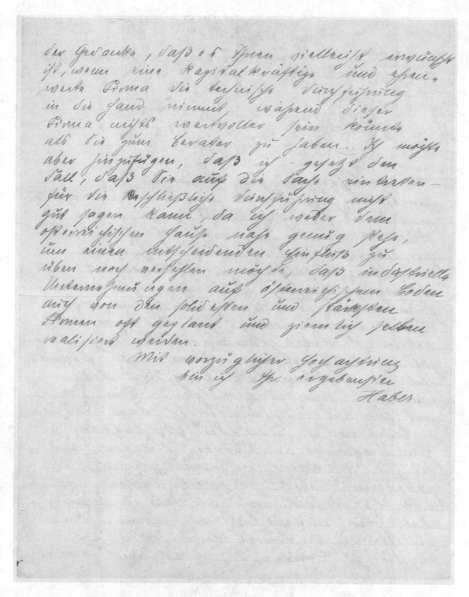

Fig. 10.5 (continued)

elements. Three reports from Haber's 1902 tour of industrial facilities in North America entitled "Über Hochschulunterricht und elektrochemische Technik in den Vereinigten Staaten" (On higher education and electrochemical technology in the United States) illustrate this view (Haber 1903a,b,d).

The busy New York harbor impressed Haber immediately upon his arrival, and the economic activity and chemical industry he saw there remained the source

Fig. 10.6 Lower Manhattan and the Brooklyn Bridge as Fritz Haber entered the New York Harbor in 1903. By David Vandermeulen from *Fritz Haber: L'Esprit du Temps* (Vandermeulen 2005); © David Vandermeulen/Guy Delcourt Productions

of his enthusiasm for the remainder of the trip (Fig. 10.6). He supplied detailed lists of numbers for production capacities as well as actual output and costs for a wide array of companies. At times, he also weighed the successful implementation of new industrial processes against adverse environmental effects on neighboring communities.[9] While ammonia itself received little attention, Haber's interest in the electric arc process was evident as he delved into an overview of its use in North America. One company among the many situated at Niagara Falls captured his attention in particular. "None of the Niagara Companies," he wrote

> has been of greater interest in the world than the Atmospheric Products Co., which has undertaken the task of oxidizing atmospheric nitrogen to nitric acid. As important as this company is, I cannot deny that its stage of development, compared to what was discussed in the journals, was a bit of a disappointment. I expected to find an operation and was confronted with an experimental apparatus, which is of course of very great interest but does not yet, in my opinion, allow a reliable prediction of technical outcome.[10]

Haber remarked on the idiosyncrasies of the company's electric arc process and compared it to other established methods. In contrast to many of the other sites he visited, here he discussed the application of the local product in addition to economic viability:

> The fact that the Atmospheric Products Co. wants to undertake a large scale implementation allows one to conclude that they are confident of profitability with today's prices for the raw materials [needed for their] current nitric acid production. If...it is true that the wear on their electrodes is quite negligible, the profitability will depend solely on the costs of power and the costs of the production of nitric acid from these dilute nitrous gases. If it were possible [i.e. were to become feasible] to fertilize with a mixture of calcium nitrite and calcium nitrate instead of with Chile saltpeter, the absorption of the dilute nitrous vapor via limestone and the use of this mixture in agriculture would offer the greatest of hopes. For 95% [concentrated] nitric acid costs 24 to 25 marks per 100 kg while that amount of calcium carbonate and calcium nitrate would presumably be cheaper to manufacture [...] The main problem is to make the saltpeter mines in the Atacama [Desert in Chile], which

---

[9] While visiting a sodium hydroxide and chlorine site near Niagara Falls, Haber wrote (Haber 1903b), "that it was confirmed to me by an American paper industrialist that the chlorinated lime [bleaching powder] fulfills all expectations [...] However, the chlorinated lime works of the company are also cause for concern. The complaints of the inhabitants in the city, the dead trees in the area, the white powder on the roofs of the building [where the chlorinated lime is produced] reveal the sensitive losses [of the chemical]. But these shortcomings cannot detract from findings about the electrolytic method and its value." *German:* "Dass der Chlorkalk nun allen Ansprüchen genügt, ist mir von einem amerikanischen Papierindustriellen bestätigt worden [...] Immerhin giebt der Chlorkalkbetrieb der Firma noch immer zu Bedenken Anlass. Die Klagen der Einwohner in der Stadt, die abgestorbenen Bäume in der Umgebung, die weiss bestaubten Dächer des Chlorkalkhauses verraten empfindliche Chlorkalkverluste. Aber diese Mängel können das Urteil über das elektrolytische Verfahren und seinen Wert nicht beeinträchtigen."

[10] "Keines der Niagara-Unternehmungen hat in der Welt mehr Interesse gefunden, als das der Atmospheric Products Co., welche den atmosphärischen Stickstoff zu Salpetersaure zu oxydieren unternimmt. So wichtig dieses Unternehmen ist, so kann ich nicht leugnen, dass seine Entwicklungsphase nach dem, was darüber durch die Blätter gegangen ist, mich ein wenig enttäuscht hat. Ich erwartete einen Betrieb zu finden und traf einen Versuchsapparat, welcher freilich von sehr grossem Interesse ist, nach meinem Urteil aber den technischen Ausgang noch nicht sicher voraussehen lässt."

are moving toward depletion, obsolete via a process which enables the production of nitric acid from air. After the agricultural sector, which uses by far the largest share of saltpeter production, the explosives industry is the most significant consumer.[11]

Haber was skeptical about the future upscaling and integration of nitric acid production for both the agriculture and explosives industries. Nevertheless, he praised the company:

Making concentrated nitric acid from nitrous vapor and air is an old problem in inorganic engineering which has yet to be solved satisfactorily. The electrochemical success of the Atmospheric Products Co. gives this inorganic-technical ambition increased emphasis. A mature, new industry has, therefore, not yet been developed. Nevertheless, it may not be denied here, that a highly meaningful step has been taken. . .[12]

Despite the attention given to it here, Haber devoted relatively little space to fixed nitrogen in his three reports on North American industry—rather it was the technical aspects of industry, in general, that interested him. The reports do, however, reflect Haber's attitude toward a synthetic source of fertilizer: the industry was not yet mature, but, if fully realized, it would provide a great benefit to agriculture. The potential for success depended on market realities rather than a moral imperative.

The reports on North American industry also illustrate another of Haber's beliefs at the time: the importance of continued industrialization and how Germany could stay at the forefront by remaining open to international communication and relationships. As the title of the reports indicate, his ideas were coupled strongly with the proliferation of chemistry education and research–a subject to which he remained deeply committed in different forms his entire career (Engler et al. 1909; Schlenk 1934), (Coates 1937, pp. 1663–1664, 1671–1672), (König 1954; Krassa 1955), (Stoltzenberg 1994, chapter 12), (Szöllösi-Janze 1998b, chapter 9). By 1903,

---

[11] "Der Umstand, dass die Atmospheric Products Co. eine Ausführung im grossen Maassstabe unternehmen will, lässt schliessen, dass sie bei den heutigen Preisen [der Rohstoffe] der derzeitigen Salpetersäureproduktion, eine Rentabilität fur gesichert hält. Wenn es sich. . .weiter bestätigt, dass die Abnutzung der Elektroden eine durchaus verschwindende ist, so hängt die Wirtschaftlichkeit in der That lediglich an den Kosten der Kraft und an den Kosten der Aufarbeitung dieser dünnen nitrosen Gase zu starker Salpetersaure ab. Wenn es möglich wäre, mit einem Gemenge von Calciumnitrit und Calciumnitrat statt mit Chilisalpeter [sic] zu düngen, so würde die Absorption der dünnen nitrosen Dämpfe mit Kalk und die Verwendung dieser Masse in der Landwirtschaft die grössten Hoffnungen bieten. Denn die Salpetersäure kostet in der Gestalt des 95 prozentigen Chilisalpeters 24 bis 25 Mk. pro 100 kg, während jenes Gemenge von Calciumnitrat und Calciumnitrit voraussichtlich billiger herzustellen wäre [. . .] Das grosse Problem liegt darin, die der Erschöpfung entgegengehenden Salpeterlager in [der] Atacama [Wüste in Chile] durch ein Verfahren entbehrlich zu machen, welches Salpetersäure aus der Luft zu gewinnen ermöglicht. Nächst der Landwirtschaft, die weitaus den grössten Teil der Salpetererzeugung aufnimmt, ist die Sprengstoffindustrie der bedeutendste Konsument."

[12] "Aus nitrosen Dämpfen mit Luft konzentrierte Salpetersäure zu machen, ist ein altes und bisher nie voll befriedigend gelöstes Problem der anorganischen Technik. Der elektrochemische Erfolg der Atmospheric Products Co. verleiht diesem anorganisch-technischen Desideratum vermehrten Nachdruck. Eine fertige neue Industrie liegt also noch nicht vor. Immerhin darf nicht verkannt werden, dass hier ein hoch bedeutsamer Schritt geschehen ist. . .".

it had long been a recurring issue in his correspondence. "It is simply so," he wrote to Richard Abegg in March of 1902 (Haber 1902b),

> that analysis is the foundation of education. . .where else are students to find their interest in [material] things if the professor does not have it? And he cannot have it if he is working according to a completely different philosophy [. . .] Certainly one person may do more of one thing while someone else does more of another. But one thing we must have in common is that the analytical education of the young people be fostered by the inorganic faculty through a particular interest of the teachers themselves, otherwise we will only go backward.[13]

In September of the same year, he wrote to Wilhelm Ostwald, while still in the United States, about the danger of Germany falling behind in the great discipline of its own creation (Haber 1902c).

> Physical chemistry here is by all means the leading branch of the subject. In almost all of the universities that I visited. . .the physicochemists are the young driving force which without any doubt are leaders within the department and for that reason are looked at favorably by their colleagues [. . .] Generally, organic chemistry here means nothing more than gas chemistry, or a university subject in Germany.[14]

The rest of Haber's correspondence, when not concerning private issues, often dealt with theoretical thermodynamic considerations—both in the realm of dilute solutions and other subjects. Relevant here is Haber's critique of Hans von Jüptner's derivations of the free energies of chemical reactions in the *Zeitschrift für anorganische Chemie* in 1904 because determinations of the free energy were at the heart of Haber's interest in ammonia synthesis (Haber 1904c; von Jüptner 1904a,b,c,d). However, although Haber and his assistant Gabriel van Oordt were already at work on ammonia at this time, no discussion of it is found in his letters. Nowhere is there evidence that Haber had more of an interest in a synthetic source of ammonia than would be expected from any physical chemist with close ties to industry at the beginning of the twentieth century. If anything, he was of the opinion that ammonia synthesis from the elements was not feasible on an industrial scale. Both in his 1903 report on his trip to North America and in his 1920 Nobel lecture his method of choice was clear: "No better and more economical process for the binding of nitrogen could. . .be devised if some means could be found for converting

---

[13] "Es ist doch nun einmal so, daß Analysen des Unterrichts Fundament sind. . .wo soll das Interesse an diesen [stofflichen] Dingen beim Student herkommen, wenn der Professor es nicht hat und der kann's nicht haben, wenn er ganz anderer geistiger Arbeitsrichtung folgt [. . .] Gewiss soll der eine mehr dies, der andere mehr jenes treiben. Aber gemeinsam muß bleiben, daß die analytische Bildung der jungen Leute von den anorganischen Lehrstellen aus durch ein besonderes Interesse der Lehrenden gefördert wird, sonst gehen wir immer wieder zurück."

[14] "Die physicalische Chemie ist hier schlechterdings der führende Zweig der Faches. In fast allen Universtäten, die ich besucht habe. . .sind physicochemiker diejenigen jüngeren Kräfte, die ganz unzweifelhaft wissenschaftlich innerhalb der Facultät führen und dafür auch von den anderen Fachgenossen angesehen werden [. . .] Generell bedeutet organische Chemie hier nicht mehr als Gaschemie, als Lehrfach in Deutschland."

electrical energy into. . .chemical energy [via the electric arc] without waste (Haber 1920)."

Based on the information presented here and in the following chapters it will become evident that Haber's scientific attraction to ammonia synthesis from the elements—which he and others had described as fundamental research—was based on his interest in the free energies of chemical reactions. The equilibrium between ammonia, hydrogen, and nitrogen was an optimal system to test the theoretical work he had done, for example, in his book *Thermodynamik technischer Gasreaktionen* (Haber 1905b). It was not simply a new application of old concepts, but rather, as quoted above, "a scientific confirmation of basic principles"—just the kind of question Haber had pursued during his transition toward topics of increasing fundamental significance. Because it was basic research, it was not mandatory to have investigated ammonia equilibrium at all—there were any number of gas systems to choose from, but ammonia was especially interesting. Haber's task was not only to qualitatively solve the old "riddle of ammonia synthesis," that is, that ammonia could be easily catalytically dissociated but not generated (see Part I, Chap. 4). Instead, he could apply the new principles of physical chemistry to demonstrate precisely why ammonia generation had alluded researchers for so long. Furthermore, there were many studies already available in the literature containing important thermal data on the ammonia system. Apart from the science, a financial aspect was also present: the industrialist Margulies brothers incentivized Haber's choice with financial compensation and funding for research as is often the case today (Ertl 2018). These factors become especially conspicuous when considering Haber had no pronounced moral impulse at the time to target the synthesis of fixed nitrogen in any form. When he started the work, he also began studies of the multistep process and the electric arc, although he was not convinced these could be made economical either (Haber 1905b, pp. 189–190). If the Margulies brothers had offered him financial incentives to investigate another chemical system, he may very well have accepted that task instead–but they did not. They asked him to consider the ammonia system because nitrogen fixation had economic potential for industry. At this point in his career, having been made a full professor five years earlier, Haber had the required security in his career to take a risk he thought had little chance of success apart from increased basic scientific knowledge.

Chapter 11
The Scientific Publications on Ammonia Synthesis

Having set the stage, we now turn to the analysis of Haber's and Nernst's published results between 1905 and 1908. A short commentary is in order here—both on style and on the use of theory. In summarizing the articles to emphasize pertinent scientific content, it was not possible to express the tone of the century-old publications and the interested reader is encouraged to review the original sources. As is occasionally mentioned, the number of calculation errors is noticeable as is the obsessiveness with which Haber, Gabriel van Oordt, Robert Le Rossignol, and later Friedrich Jost considered qualitative and quantitative sources of experimental error. The publications are filled with details about uncertainties, insecurities, and checks on sources of inaccuracy that remind today's reader that these experiments were conducted by human researchers at the limits of then-current science. One characteristic examples is from Haber and Le Rossignol's 1908 publication (Haber and Rossignol 1908f). During stability tests for the iron catalyst at 30 atm, the nitrogen-hydrogen gas mixture began to run low and further tests had to be performed at 22.35 atm (this lower pressure changed the gas flux and they were concerned with the effect on equilibrium). The results were compared with the help of theoretical corrections and full details were supplied—including the fact that the effect of the gas flux on the catalyst's effectivity could not be calculated. It was a description of real laboratory difficulty and was not altered to make the study appear more structurally sound than it was. In contrast, today's publications are often polished in such a way as to fend off attack (also, today one would simply purchase new gas cylinders with the required pressure).

As for theory, the fitting functions found here are key to Haber's and Nernst's analyses and extrapolation of the behavior of the ammonia system at temperatures different from their chosen experimental conditions. For those who are interested, the derivation of these functions can be found in Appendix A and a careful examination, along with Appendix B, will provide a deeper understanding of the then-current physicochemical approach to chemical equilibrium, especially with respect to thermal data. The equations are also compared to modern calculations

© The Author(s) 2022
B. Johnson, *Making Ammonia*, https://doi.org/10.1007/978-3-030-85532-1_11

of ammonia equilibrium in Figs. 11.7, 11.8, 11.9, 11.10, and 11.11.[1] Those not inclined to consult the appendices may simply read on!

11.1   Über die Bildung von Ammoniak aus den Elementen (On the Generation of Ammonia from the Elements) by Fritz Haber and Gabriel van Oordt, 1905

According to physical chemists in 1905, the most promising method for synthesizing fixed nitrogen was expected to be the electric arc, while coking operations would continue to generate meaningful contributions (Ostwald 1903). However, the uncertainty about the effectiveness of these methods meant that the multi-step process and direct synthesis from the elements were simultaneously pursued. In that year alone, Haber himself investigated all three. The production of nitric oxide using the electric arc had become an ongoing project with Adolf König (Haber 1905b, pp. 86–89), (Haber and König 1907d, 1908e; Haber et al. 1910) while with Gabriel van Oordt he published the first accurate, quantitative study on ammonia synthesis from the elements describing their work for the Margulies brothers (Haber and van Oordt 1905b).[2] The publication also contained the results of their investigation of the multi-step process based on the work of Moissan and Güntz. Several years earlier, these researchers had cycled calcium and barium nitride in H_2 atmosphere to calcium and barium hydride in order to catalytically synthesize ammonia (Güntz 1898; Moissan 1898). Haber and van Oordt used the metals calcium and manganese. With calcium nitride they obtained ammonia only with the reaction $Ca_3N_2 + 6H_2 \rightarrow 3CaH_2 + 2NH_3$; cycling back to nitride via the reaction $3CaH_2 + 2N_2 \rightarrow Ca_3N_2 + 2NH_3$ had no result, as expected. According to theory, ammonia yields by this process should remain below the detection limit. Reactions with manganese were similar although the minimum working temperature was lower. Nevertheless, the reaction rates were too small to be of interest, and the cycling experiments were abandoned. Although today the results for direct ammonia synthesis are considered to be the scientifically and historically significant parts of this publication, at the time, the multi-step process was prominent and took up the entire introduction of the article (Fig. 11.1).

After the multistep process, Haber and van Oordt described their results for the direct production of ammonia from the elements. It is important to keep in mind that for this reaction, there had not yet been any accurate determination of equilibrium. Haber and van Oordt assumed though "that the generated yield will be very small."[3] They proceeded carefully by determining the equilibrium "from both

---

[1] The modern calculations were provided by Dr. Kevin Kähler at the Max Planck Institute for Chemical Energy Conversion.

[2] These results were also reported and discussed in (Haber 1905b, pp. 185–190).

[3] "...daß die gebildete Menge sehr klein sein wird."

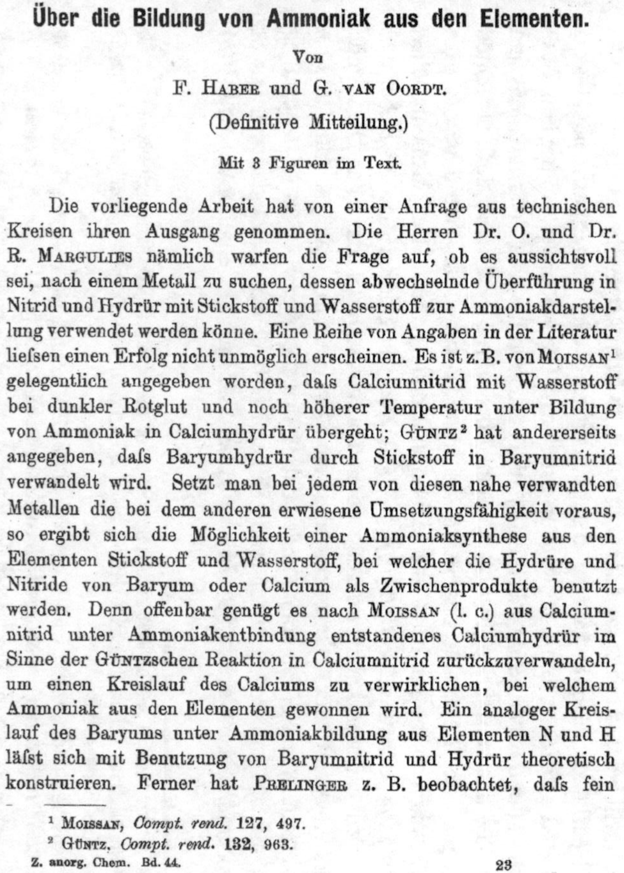

On the Production of Ammonia from the Elements.

by

F. Haber and G. van Oordt.

(Definitive Contribution.)

With 3 Figures in the Text.

The present Work was started due to an inquiry coming from the technical community. It was namely Misters Dr. O and Dr. R. Margulies who posed the question whether it were promising to search for a metal which, when cycled from nitride to hydride with nitrogen and hydrogen, could be used for the production of Ammonia. A series of studies in the literature did not seem to rule out success. It has been reported, for example, by Moissan[1], that calcium nitride reacts with hydrogen to calcium hydride and ammonia at low red heat and higher temperature; Güntz[2] on the other hand has reported that barium hydride reacts to barium nitride with nitrogen. If one assumes for each of these closely related metals the same affinity for reaction, the result is the possibility of an ammonia synthesis from the elements nitrogen and hydrogen, in which the hydrides und nitrides of barium or calcium are used as intermediate products. For it is apparently adequate, according to Moissan (l. c.), to convert the calcium hydride, which is produced from calcium nitride during the production of ammonia, back into calcium nitride in the sense of the Güntz's reaction, so that ammonia is produced from the elements. An analogous cycle of barium using barium nitride and hydride to produce ammonia from the elements N and H can be theoretically devised...

Über die Bildung von Ammoniak aus den Elementen.

Von

F. Haber und G. van Oordt.

(Definitive Mitteilung.)

Mit 3 Figuren im Text.

Die vorliegende Arbeit hat von einer Anfrage aus technischen Kreisen ihren Ausgang genommen. Die Herren Dr. O. und Dr. R. Margulies nämlich warfen die Frage auf, ob es aussichtsvoll sei, nach einem Metall zu suchen, dessen abwechselnde Überführung in Nitrid und Hydrür mit Stickstoff und Wasserstoff zur Ammoniakdarstellung verwendet werden könne. Eine Reihe von Angaben in der Literatur ließen einen Erfolg nicht unmöglich erscheinen. Es ist z.B. von Moissan[1] gelegentlich angegeben worden, daß Calciumnitrid mit Wasserstoff bei dunkler Rotglut und noch höherer Temperatur unter Bildung von Ammoniak in Calciumhydrür übergeht; Güntz[2] hat andererseits angegeben, daß Baryumhydrür durch Stickstoff in Baryumnitrid verwandelt wird. Setzt man bei jedem von diesen nahe verwandten Metallen die bei dem anderen erwiesene Umsetzungsfähigkeit voraus, so ergibt sich die Möglichkeit einer Ammoniaksynthese aus den Elementen Stickstoff und Wasserstoff, bei welcher die Hydrüre und Nitride von Baryum oder Calcium als Zwischenprodukte benutzt werden. Denn offenbar genügt es nach Moissan (l. c.) aus Calciumnitrid unter Ammoniakentbindung entstandenes Calciumhydrür im Sinne der Güntzschen Reaktion in Calciumnitrid zurückzuverwandeln, um einen Kreislauf des Calciums zu verwirklichen, bei welchem Ammoniak aus den Elementen gewonnen wird. Ein analoger Kreislauf des Baryums unter Ammoniakbildung aus Elementen N und H läßt sich mit Benutzung von Baryumnitrid und Hydrür theoretisch konstruieren. Ferner hat Perlinger z. B. beobachtet, daß fein

[1] Moissan, Compt. rend. 127, 497.
[2] Güntz, Compt. rend. 132, 963.
Z. anorg. Chem. Bd. 44.

23

Fig. 11.1 The first page of Fritz Haber and Gabriel van Oordt's 1905 Publication (and English translation) (Haber and van Oordt 1905b)

sides" ($2\,NH_3 \rightarrow 2\,N_2 + 3H_2$ and $2\,N_2 + 3H_2 \rightarrow 2\,NH_3$) using a novel apparatus with two heated, ceramic pipes containing iron and nickel catalysts.[4] Dry ammonia gas was dissociated in the first pipe, and the resulting nitrogen and hydrogen gases were passed through a known volume of sulfuric acid to remove and determine the remaining quantity of ammonia. The gases were redried and fed into the second ceramic pipe, which was held at the same temperature as the first. Here, ammonia was generated and the reacted gas again passed through a known volume of sulfuric acid. By comparing the ammonia yields in the two pipes, Haber and van Oordt hoped to rule out (or identify) any influence of the apparatus or of differing reaction rates on the formation or dissociation processes. If, as in Ostwald's experiments, nitrogen contained in the catalyst had altered the measured equilibrium position in one reaction, the two ammonia yields would have differed. Agreement in both pipes was strong evidence equilibrium had been reached. The experiments were carried out at one atmosphere as the experimental conditions were demanding enough. "Practical reasons," Haber wrote, "which need not be discussed here, caused us to work not at high pressures, but at normal pressures."[5]

A series of seven measurements (Table 11.1 and the black squares in Fig. 11.7) with the porcelain pipes, an iron catalyst, and a 1:3 N_2:H_2 gas mixture led to the determination that the equilibrium partial pressure of NH_3 at 1293 K was $p_{NH_3} = 0.12 \times 10^{-3}$ atm (0.012 Vol.-%) and is shown as the gray hexagon in Fig. 11.7. This corresponded to an equilibrium constant of

$$K_p = \frac{(p_{N_2})^{1/2}(p_{H_2})^{3/2}}{(p_{NH_3})} = \frac{0.25^{1/2}0.75^{3/2}}{0.12 \times 10^{-3}} = 2706 \qquad (11.1)$$

which will be used below together with Eq. A.23 from Appendix A.

Experiments I and II were carried out in a different oven than experiments III–VII and used a different iron catalyst (iron asbestos). The amount of ammonia in both pipes in these experiments was notably higher than in the later determinations and indicated that equilibrium had been reached but with a systematic shift, attributed to a possible irreversible change in the iron. Haber wrote, "we therefore give little weight to the results of the second and mainly the first experiments when calculating a mean value [for p_{NH_3}] (Haber and van Oordt 1905b)."[6] Haber and van Oordt had assessed the time dependence of the catalyst's activity and had found none. However, later experiments confirmed the fresh iron catalyst had caused a short term excess of ammonia, underscoring that the overall effect of the catalyst was still not completely understood (Haber 1920). It was concluded that out of 1000 ammonia molecules 999.76 would be dissociated at equilibrium with 0.24

---

[4] The importance of the experimental apparatus for the scientific breakthrough should not be understated. The technical details remain brief here with more complete descriptions found in the original publications.

[5] "Praktische Gründe, welche wohl keiner besonderen Erläuterung bedürfen, veranlaßten uns, nicht bei hohen Drucken, sondern bei gewöhnlichem atmosphärischen [sic] Drucke zu arbeiten."

[6] "Wir legen deshalb den Ergebnissen des zweiten und vor allem des ersten Versuches bei der Bildung eines Mittelwertes [für p_{NH_3}] nur geringes Gewicht bei."

Table 11.1 Values given by Haber and van Oordt for ammonia formation/dissociation experiments over iron and a 1:3 N_2:H_2 gas mixture. Experiments VIII* and IX* are shown for later reference and were not included in the table in 1905. The values appear as black squares in Fig. 11.7

Experiment number	NH_3 pro mille that **did not** dissociate (pipe 1)	NH_3 pro mille formed (pipe 2)	Catalyst
I	0.98	0.98	Fe
II	0.46	0.46	Fe
III	0.20	0.26	Fe
IV	0.21	0.14	Fe
V	0.23	0.16	Fe
VI	0.15	0.14	Fe
VII	0.20	0.21	Fe
VIII*	0.25	0.11	Ni
IX*	0.485	0.272	Ni

remaining, a number higher than any result in experiments III through VII except pipe 2 in experiment III but far below the values in experiments I and II.

Experiments VIII and IX were made using a nickel catalyst (nickel nitrate reduced with hydrogen) but were not originally considered accurate assessments of equilibrium and were not included in the consideration of the final equilibrium value. They were performed in the same apparatus as experiments III through VII, but in both cases the amount of ammonia found in pipe 1 (dissociation) was more than in pipe 2 (generation). This was due, wrote Haber, "to the low catalytic activity [of nickel] compared to iron so that equilibrium was only just reached."[7] We will return to these measurements later.

To graph the results of the iron experiments, Haber and van Oordt used Eq. A.23 for the free energy. They chose a heat of formation for ammonia of 12000 cal/mol at standard temperature and constant pressure, so that the value at 0 K would be 10329 cal/mol. It was a difficult determination according to Haber because the mean specific heat for ammonia was still uncertain. Eq. A.23 became

$$A = 10329 - 14.21\,T\,(log\,T - 1.35) + 0.0014\,T^2 + 4.56\,T\,log\,\frac{(p_{N_2})^{1/2}(p_{H_2})^{3/2}}{p_{NH_3}}$$

(11.2)

where the final term was the experimentally determined fraction of the partial pressures, or equilibrium constant $K_p = 2706$, from Eq. 11.1. By setting $A = 0$ Haber could determine the thermodynamic equilibrium of a mixture of N_2, H_2, and NH_3 at any temperature and calculated the numbers in Table 11.2 (Fig. 11.7, solid line).

---

[7] "Dies liegt daran, daß Nickel eine viel geringere katalytische Wirkung übt als das Eisen und daß das Gleichgewicht deshalb nur knapp erreicht wird."

However, Haber made a mistake. If the values in Table 11.2 had been calculated with Eq. 11.2 using the integration constant determined with the measured value at 1293 K, this point would lie exactly on the curve, but it does not. The rest of the points from 27 °C to 927 °C do; it was a mathematical error and the volume of ammonia in equilibrium was actually lower than that calculated with Eq. 11.2.[8] The correct function is shown as the dashed line in Fig. 11.7.

After these first measurements, Haber was convinced his results were of sufficient accuracy, but that they were not promising for industry. His skepticism was tied not only to the unfavorable equilibrium position, but also to the efficacy of the catalyst (Haber and van Oordt 1905b):

> It follows further from our table [Table 11.2] that from the beginning of red heat onward, no catalyst can produce more than traces of ammonia at normal pressure with the most favorable gas mixture. Even with much increased pressure, the equilibrium position remains unfavorable.[9]

It was the obdurate di-nitrogen which was the problem. "This peculiar inertness of nitrogen," he wrote in his 1905 textbook (Haber 1905b, pp. 189–190),

> will always pose economical difficulties for a technical production of ammonia from the elements especially because nature offers us huge quantities of organically bound nitrogen in coal that is easily converted to ammonia.[10]

In addition to their main publication (definitive Mitteilung) just discussed (Haber and van Oordt 1905b), Haber and van Oordt also published a preliminary letter (vorläufige Mitteilung) (Haber and van Oordt 1905d) earlier in 1905, after becoming

Table 11.2 Calculated values from Eq. 11.2 given by Haber and van Oordt in 1905 for the thermodynamic equilibrium of a mixture of N_2, H_2, and NH_3 at 1 atm, and at different temperatures. Only the point at 1020°C was measured.

Temp [°C]	Vol.-% H_2	Vol.-% N_2	Vol.-% NH_3
27	1.12	0.37	98.51
327	68.46	22.82	8.72
627	74.84	24.95	0.21
927	75	25	0.024
1020	75	25	0.012

---

[8] There is no indication the error effected Haber's attitude toward ammonia synthesis. He was unlikely more pessimistic than he otherwise would have been because the error made the equilibrium position appear more advantageous.

[9] "Weiter aber folgt aus unserer Tabelle [Tabelle 11.2], daß von beginnender Rotglut aufwärts kein Katalysator mehr als Spuren Ammoniak bei der günstigsten Gasmischung erzeugen kann, wenn man bei gewöhnlichem Drucke arbeitet. Auch bei stark erhöhtem Druck bliebe die Lage des Gleichgewichts stets eine ungünstige."

[10] "Diese eigentümliche Stickstoffträgheit wird einer technischen Ammoniakdarstellung aus den Elementen besonders darum immer wirtschaftlich Schwierigkeiten bereiten, weil die Natur uns ungeheuere Massen organisch gebundenen und leicht in Ammoniak überführbaren Stickstoffs in der Kohle zu Gebote stellt."

aware of Edgar Perman's work from the previous year (Perman 1904a; Perman and Atkinson 1904a,b). Perman's research consisted of a series of dissociation experiments in a porcelain container, described as being a continuation of the work of Ramsay and Young (it was William Ramsay himself who read the paper before the Royal Society in London). However, despite knowledge of Ostwald's ideas, Perman's approach was empirically based, and he made no attempt to use modern physicochemical concepts. This point was emphasized by the initial referee William Cecil Whethan who did not recommend the paper for publication (Whetham 1904).

"We have confirmed," wrote Perman, for example,

> the observations of Ramsay and Young that the decomposition is never absolutely complete, but after heating at a temperature of $1100°$ for a short time the amount of ammonia remaining is so small that it can be neglected for practical purposes.

Perman believed the generation of ammonia to be reversible but decomposition to be irreversible. That is, there was "no indication whatever of any. . .equilibrium."

Haber and van Oordt's preliminary letter was a summary of the data and discussion contained in their main publication—establishing them in the field of ammonia synthesis—and though at odds with their own, they did not consider Perman's results or methodology beyond a simple acknowledgment. In 1905, Perman published again with increased attention to synthesis reactions, including the electric arc and multi-step processes (Perman 1905). The paper also required revisions before publication but was received more warmly than the previous one (Scott 1905). However, apart from a more detailed look at ammonia generation with various catalyst materials, it continued Perman's empirical approach and disregard of the principles of physical chemistry (Perman 1904a, 1905; Perman and Atkinson 1904a; Perman and Davies 1906). Traces of ammonia were only found after heating a mixture of nitrogen and hydrogen with an iron catalyst in the presence of moisture. "The proportion of ammonia [in Haber and van Oordt's experiments]," Perman wrote,

> . . .was about 0.2 to 1000. . .which is considerably less than obtained by me [. . .] Moreover, the part played by the iron is not yet completely understood. My experiments show that the quantity of ammonia formed depends on the amount of moisture present, but Haber and van Oordt appear to have overlooked this point and simply say that their gases were dry.

In their response, Haber and van Oordt now took a position in regard to Perman's work (Haber and van Oordt 1905f). They defended their conclusions and drew attention to Perman's inconsistencies—especially in relation to the role of moisture and his apparent lack of understanding of physical chemistry. "We hope," wrote Haber, "that Herr Perman, by giving some attention to the application of thermodynamics to chemical equilibrium, will come to a different consideration."[11] Despite some experimental uncertainty, Haber was confident in his use of these

---

[11] ". . .wir geben uns der Hoffnung hin, daß Herr Perman bei einiger Beschäftigung mit den Anwendungen der Thermodynamik auf chemische Gleichgewichte zu einer anderen Einsicht gelangen wird."

principles and the order of magnitude of the results he had obtained with van Oordt. Further evidence is contained in a letter to Richard Lorentz, editor of *Zeitschrift für Elektrochemie*, signaling the impending response to Perman's 1905 article (Haber 1905a):

> Two pages will be sent to you from van Oordt, which I request to be included in your journal. I am sorry that this time I am only sending a defense against an unneeded attack. Indeed, I promise something better for next month.[12]

It is the first mention of ammonia synthesis in Haber's surviving correspondence and he supplied no further scientific detail on the subject. Instead, he turned to the success he had had in his ongoing study of the carbon element (Haber and Moser 1905c; Haber and Bruner 1904g). This lack of comment to his colleagues on the ammonia system was in stark contrast to other topics such as dilute solutions.

The exchange in literature with Perman was not unique for Haber and illustrates the circumstances surrounding the equilibrium of ammonia in 1905. While there was widespread agreement that the effect of the catalyst was not fully understood, the state of application of physicochemical principles was more ambiguous. Despite their demonstrated success, the existence of equilibrium and the reversibility of chemical reactions were not universally accepted. Discrepancies even existed amongst close colleagues and renowned researchers (Le Rossignol, for example, had worked with Ramsay in England and joined Haber by 1906 (Jaenicke 1959)). Haber and van Oordt's experimental skill and intuition also become evident here. While Ramsay and Young, Ostwald, and Perman were not prepared (or able) to consistently measure (or seriously consider) such minute quantities of ammonia (see Figs. 9.3, 9.4, and 9.5), Haber and van Oordt did so on the first attempt in a way that was remarkably accurate for the tools at hand: compare the modern theoretical curve to Haber's data points in Fig. 11.7 while remembering that Nernst's calculations were not yet available. *Up until this point, no one knew–not even to an order of magnitude–how much ammonia would be in equilibrium with nitrogen and hydrogen.*

At the time, however, the theoretical confirmation of Haber's measurements was underway. It was a result that would unequivocally establish the authority of physical chemistry in solving the mystery of ammonia synthesis.

---

[12] "Durch van Oordt werden Ihnen 2 Seiten zugesan [sic], die ich in Ihre Zeitschrift aufzunehmen bitte. Es thut mir leid, daß es bloß eine Verteidigung gegen eine [sic] unnötigen Angriff ist, die ich diesmal zu senden habe. Doch verspreche ich für nächsten Monat besseres."

11.1.1 Über die Berechnung chemischer Gleichgewichte aus thermischen Messungen (On the Calculation of Chemical Equilibrium from Thermal Measurements) by Walther Nernst, 1906

Walther Nernst's initial publication on the subject of ammonia synthesis grew out of theoretical work from the second half of the nineteenth century, which investigated the nature of the heat released during chemical reactions—work that established the existence and role of the free energy in chemical dynamics (Figs. 11.2 and 11.3). At the beginning of the twentieth century, this theoretical framework, while having provided pivotal information on the nature of a chemical reaction, was not able to give a complete quantitative picture. A complementary experimental measurement was needed to calculate the equilibrium of chemical systems, including ammonia. Finding a complete and general theoretical treatment had become one of the central problems in physical chemistry. While Haber and others developed methods for employing the experimental solution (see Appendix A), Nernst continued to consider the theoretical approach. He began with the work of Hermann von Helmholtz and, using a mathematical equation that would later be called the third law of thermodynamics, found a solution that allowed the calculation of thermodynamic equilibria using only the results of thermal data from vapor pressure measurements. At first, however, it was much less dramatic: Nernst saw his new equation as little more than a mathematical tool obtained from a hunch after reviewing experimental data. This development is outlined in more detail in Chap. 13 and Nernst's full derivation can be found in Appendix B.

Here we concentrate on the sections of Nernst's work devoted to the calculation of the ammonia system as it related to Haber's research.

In the second section of his paper (the first section is the derivation of his heat theorem found in Appendix B), Nernst focused on empirical and theoretical methods of obtaining and evaluating vapor pressure curves with which he determined the unknown constant of integration. While this section contained physicochemical considerations, for example, the numerical values and role of molar specific heats in calculations, it was mainly an assessment of the inadequate state of vapor pressure data–ammonia being an important exception (Dieterici 1904). He also gave a table of chemical constants, including those needed to calculate the equilibrium of ammonia, which became central to the interaction with Haber: H_2 had a value of 2.2, N_2 had the value of 2.6, and NH_3 had the value 3.3. Added together they gave the value of the integration constant, equal to the term Σvi in Eq. B.32.

In the third section of his paper, Nernst applied these data and his heat theorem to several examples. For the $N_2 + 3H_2 \leftrightarrow 2NH_3$ system, he obtained the following expression for the logarithm of the equilibrium constant K:

Fig. 11.2 Walther Nernst circa 1906. *Source*: A digital copy of this photo was obtained from the Archive of the Max Planck Society, Berlin-Dahlem–its origin is unknown

Fig. 11.3 Portrait of Nernst in 1914 by David Vandermeulen from *Fritz Haber: Les Héros* (Vandermeulen 2007); ©David Vandermeulen/Guy Delcourt Productions

$$log\ K = log\ \frac{p_{H_2}{}^3 \cdot p_{N_2}}{p_{NH_3}{}^2} = log\ \frac{0.75^3 \cdot 0.25}{x^2} = -\frac{24000}{4.571\ T} + 3.5\ log\ T + 6, 6 + 2, 6 - 6, 6$$

(11.3)

where the factor 4.571 is obtained as in Eq. A.27 and the three factors at the end of the equation are the chemical constants for H_2 (2.2), N_2 (2.6), and NH_3 (3.3)–there are three moles of molecular hydrogen and one mole of molecular nitrogen in two

moles of ammonia.[13] When referring to "thermal data" later, it will be in reference
to these numbers along with the specific heat and the heat of formation of ammonia.

Equation 11.3 reduced to:

$$log \frac{0.325}{x} = -\frac{2625}{T} + 1.75 \, log \, T + 1.3 \tag{11.4}$$

and is shown as the solid line in Fig. 11.8.

Using Haber and van Oordt's value for the partial pressure of ammonia from 1905
(Haber and van Oordt 1905b), $x = 0.12 \cdot 10^{-3}$, Nernst calculated a corresponding
absolute temperature of 893 K, whereas Haber and van Oordt had reported a
temperature of 1293 K–a difference of 400 degrees (Fig. 11.8, black squares).[14]
Nernst made no further comment at the time. Instead, he offered a final thought on
his calculations for homogeneous gas phase systems. Although he felt the numerical
results of his theory spoke for themselves, he hoped it would not be the only useful
aspect of his work. "Of greater importance," Nernst wrote,

> is the fact that the general character of the relationship between chemical energy and heat
> seems, after a long search, to unveil itself from the fundamental theorem in harmonic
> simplicity.[15]

The third law delivered information not contained in the first and second laws and
allowed Nernst to move beyond Helmholtz. He was able to determine the constant of
integration which was tantamount to disentangling the total energy from the entropy,
bound together by their mutual definition (Eq. B.3). It was also the last step to a
full theoretical formulation of classical thermodynamics (see Chap. 13) (Haberditzl
1960; Suhling 1972), (Müller 2007, pp. 171–172).

11.1.2 Über das Ammoniak-Gleichgewicht (On Ammonia Equilibrium) by Fritz Haber and Robert Le Rossignol, 1907

In connection with the results from his heat theorem and supporting experiments,
Walther Nernst wrote a letter to Fritz Haber in the fall of 1906 concerning
Haber's 1905 publication. Working at lower temperatures and higher pressures,

---

[13] This equation was essentially the same as Eq. B.21, except that here Nernst used $Q = Q_o = constant = 24,000$ cal (Eq. A.25) for the heat of formation of two moles of ammonia. The choice has no effect on the determination of the integration constant but is interesting considering the earlier use of the more accurate expression for Q in Eq. A.22.

[14] Nernst's equation actually results in a temperature of 871 K for Haber's value of x.

[15] "Wichtiger scheint mir der Umstand, daß der allgemeine Charakter der Beziehung zwischen chemischer Energie und Wärme durch die Formeln, welche aus dem Fundamentaltheorem fließen, sich nach langem Suchen in harmonischer Einfachheit zu enthüllen scheint."

Nernst, together with Karl Jellinek, had found ammonia yields three to four times smaller than expected from Haber's extrapolations. Haber's determination of the equilibrium was apparently incorrect. Although Haber was skeptical that Nernst had actually measured the equilibrium state (according to Haber, Nernst had only performed ammonia generation experiments, which would result in lower yields if complete equilibrium had not been reached), he began new experiments to verify his 1905 results, this time with Robert Le Rossignol. The results were published in 1907 (Haber and Rossignol 1907e). The apparatus remained largely the same, with several improvements including a more accurate temperature assessment at the exact site of the reaction[16]. During their work, Walther Nernst published his heat theorem, which, along with the earlier letter, changed Haber's perspective and approach to ammonia synthesis. It was in 1906 and early 1907 (and *not* at the Bunsen Society meeting in May 1907) that Haber began to entertain the idea of a range of values resulting from his 1905 experiments and that one of these—in fact one of the lower values he had not yet considered—was correct and could provide the necessary response to Nernst's criticism. For Haber, his determination of the order of magnitude using a weighted average near the upper bound of his earlier results had originally been sufficient—and it had been no small achievement. It was the first determination of the equilibrium of ammonia that could be viewed as a rigorous and dependable establishment of an approximate value, albeit one that indicated little promise for industrial application. But after Nernst's reaction, Haber decided to take another look (Jost 1908a). In the context of his new experiments with Le Rossignol, he repackaged his earlier results to draw attention to any agreement between the 1905 and 1907 data sets as well as to the improvements in experimental accuracy.

Referring to the ammonia yields in pipes 1 and 2 in the 1905 experiments (Table 11.1), Haber described the way in which measurements I and II, and now measurement IX with nickel (a measurement excluded in 1905) had been incorporated:

> They were given weight in such a way that rather than a mean value from the results of the six[17] other experiments, a value of 2.4 mols [per 10,000 or 0.012 Vol.-% at 1293 K] close to their upper limit was seen as correct and used for subsequent calculations.[18]

This statement may be compared to the quote in 1905 on page 110 on how the average was obtained. Not only did Haber now consider the nickel experiments (VIII and IX in Table 11.1) to be significant, he also began to refer to an upper limit within a range of values. By including the nickel experiments he was able to substantiate

---

[16] Images of Haber's laboratories and staff around the time of his work on ammonia synthesis are shown in Figs. 11.4, 11.5, and 11.6.

[17] Here, Haber was referring to the five iron experiments, III-VII and the first nickel experiment (VIII in Table 11.1).

[18] "Ihnen [Versuchen I und II in Tabelle 11.1] wurde insofern Rechnung getragen, als nicht das Mittel aus dem Ergebnis der sechs anderen Versuche, sonder ein deren obere Grenze naheliegender Wert von 2.4 Molen für richtig genommen und der weiteren Rechnung zugrunde gelegt wurde."

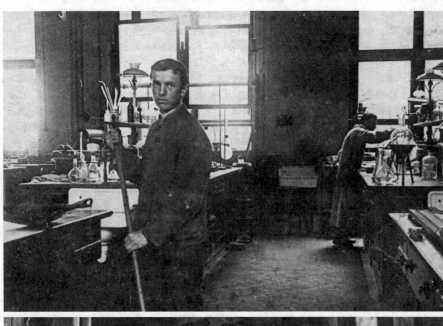

Fig. 11.4 Members of Haber's staff at the Institute for Physical Chemistry and Electrochemistry at the Technical University of Karlsruhe in 1909. Above: laboratory assistant (Laboratoriumsdiener) Adolf; Below: Friedrich Kirchenbauer. *Source*: Archive of the Max Planck Society, Berlin-Dahlem, Jaenicke and Krassa Collections, Picture Numbers VII/9 and VII/11

Fig. 11.5 Above: H.J. Hodsman; Below: Setsuru Tamaru. The photos also show the state of the laboratories and machinery around the time of the breakthrough of ammonia synthesis. *Source*: Archive of the Max Planck Society, Berlin-Dahlem, Jaenicke and Krassa Collections, Picture Numbers VII/12 and VII/16

Fig. 11.6 Haber and staff at the Institute for Physical Chemistry and Electrochemistry at the Technical University of Karlsruhe during the summer semester of 1908. Robert Le Rossignol sits behind and to Haber's right, in front of him is Friedrich Kirchenbauer, J.E. Coates stands in the middle row at left, to his left is H.J. Hodsman, and Paul Krassa sits in the front row, third from left. *Source:* Archive of the Max Planck Society, Berlin-Dahlem, Jaenicke Collection, Picture

the value of 0.24 mols/1000 chosen in 1905, which lay close to the upper limit of the data. Perhaps more importantly though, experiment VIII provided him with a new low value which corresponded precisely with his 1907 measurements. The upper and lower limits, stressed later at the 1907 Bunsen Society meeting (Nernst 1907), were 0.013 and 0.0055 Vol.-%, respectively, and are shown in Fig. 11.7 by the gray triangles.[19]

Haber then turned to the 1905 iron stability tests in which large volumes of gas were passed over the catalyst. It had been an effort to show the reliability of experiments III–VII in Table 11.1 because some fluctuation with time had been observed. The stability experiments showed no catalyst degradation although the equilibrium yields were less than half of those found in experiments III–VII. In light of the fact that the stability tests were performed with high gas fluxes, Haber had found the deviation in 1905 to be "not very significant,"[20]—important was that the order of magnitude was the same. He reacted differently in 1907 (Haber and Rossignol 1907e):

> The results of a longer [stability] experiment...were taken to show the ammonia yield stayed well below equilibrium due to the quick flow of the gases over the [iron] catalyst. Our new experiments have shown us that the ammonia generation in that stability test was only slightly lower than the [real] equilibrium value, which was not at the upper, but rather at the lower limit of the range in which the results of the six main experiments [III–VII and now VIII]...were found.[21]

The stability test equilibrium value is given in Fig. 11.7 as a gray star along with Eq. 11.2, recalculated using this point to determine the integration constant (dotted line).

Instead of a range of values, Haber and van Oordt had chosen originally and in less than transparent fashion to represent all of the 1905 experiments with a single value close to the "upper limit." The true equilibrium value was indeed contained in those results–it was only a matter of recasting them as a range instead of a weighted average. Haber attributed the remaining deviation reported in Nernst's letter to an old problem: the limited knowledge of the specific heat of ammonia at high temperatures.

In 1907, Haber still found the equilibrium position of ammonia unfavorable for industrial synthesis but "at lower temperatures, it was not so badly shifted as it

---

[19] The upper limit of 0.013 Vol.-% was from experiment III (Table 11.1, formation column) and the lower limit of 0.0055 Vol.-% was apparently from experiment VIII with the nickel catalyst (Table 11.1, formation column), although the reported value in 1907 at the Bunsen Society meeting was 0.0052 Vol.-%

[20] "...nicht sehr erheblich."

[21] "Das Ergebnis eines länger ausgedehnten [Stabilitäts-] Versuches...wurde dahin verstanden, daß die Ammoniakausbeute wegen raschen Leitens der Gase über den [Eisen-] Katalysator erheblich hinter dem Gleichgewicht zurückblieb. Unsere neuen Versuche lehren, daß die Ammoni- akbildung bei jenem Dauerversuch nur wenig niedriger als der Gleichgewichtswert war, der nicht an der obere, sonder an der unteren Grenze des Spielraumes gelegen ist, in welchem sich die Ergebnisse der sechs Hauptversuche...bewegen."

Table 11.3 Average values given by Haber and Le Rossignol for the equilibrium constant K_p with a mixture of N_2, H_2, and NH_3 (N_2:H_2 = 1:3) at the given temperatures and iron, nickel, chromium, and manganese catalysts (Fig. 11.9, black squares). The corresponding values for volume percent have been added for comparison with the 1905 results

°C	1000	930	850	800	750	700
$K_p \times 10^4$	1.48	2.00	2.79	3.34	4.68	ca. 6.8
NH_3 vol.-%	0.0048	0.0065	0.0091	0.0109	0.0152	~0.0221

had seemed after the older calculations [from 1905]."[22,23] As for Edgar Perman's claim that there was no equilibrium state at all that contained ammonia, Haber now dismissed it altogether. Confidence in the accuracy and interpretation of his 1907 measurements was clearly expressed in the publication.

Haber and Le Rossignol presented the results of at least 49 individual determinations of equilibrium with iron, nickel, chromium, and manganese catalysts, and a 1:3 N_2:H_2 mixture along with a table (Table 11.3) containing the mean values for $K_p = \frac{p_{NH_3}}{p_{N_2}^{1/2} p_{H_2}^{3/2}}$, the reciprocal value of that given in 1905 (Fig. 11.9, black squares). There was a shift compared to the 1905 curve in the direction Nernst had expected: the value at ~1000 °C (~1273 K) was reduced from p_{NH_3} =0.012 Vol.-% to 0.0048 Vol.-% (compare Figs. 11.7 and 11.9).

Haber followed his results by renewed theoretical considerations. Applying the approximation from Eq. A.27, he used the function:

$$log K_p = \frac{12000}{4.57\ T} - 5.89 \tag{11.5}$$

to fit his data (Fig. 11.9, solid line). 12000 was the heat of formation for one mole of ammonia and, according to Haber, was a reliable value responsible for the quality of the fit. "A complete thermodynamic treatment,"[24] however, was not possible because the heat of formation deviated from this value at higher temperatures and reliable numbers had not yet been obtained. In this context, Haber also consulted Nernst's heat theorem and fit his data with the equation used by Nernst for the ammonia system in 1906 (Eq. 11.4). He had to modify the original expression through the addition of the linear term +0.000651 T and reformulated it for one mole of ammonia (Fig. 11.9, dashed line):[25]

$$log K_p = \frac{12000}{4.571\ T} - 1.75\ log\ T - 1.3 + 0.000651\ T \tag{11.6}$$

---

[22] "...mit fallender Temperatur nicht so ganz stark verschoben wird, wie es nach der älteren Rechnung erschien."

[23] Comparison of Figs. 11.7 and 11.9 does not substantiate this behavior but illustrates Haber's improved attitude.

[24] "Eine vollkommene thermodynamische Behandlung...".

[25] The opposite signs to those used by Nernst are due to the mutually reciprocal values of K_p.

Although the resulting curve matched Haber's data, the term linear in T implied a heat of formation for ammonia (see the expression for $Q_{p,T}$, Eq. A.22) that rose more quickly with temperature than had been determined experimentally. Haber did not comment further on the heat theorem. Criticism of Nernst's work came only after Haber received more information at the 14th Meeting of the Bunsen Society in what was a pivotal act in the scientific development of ammonia synthesis from the elements.

11.1.3 Über das Ammoniakgleichgewicht (On Ammonia Equilibrium) by Walther Nernst with Experiments from Friedrich Jost, 1907

Nernst's 1907 published response to Haber and van Oordt's 1905 study (their 1907 contribution appeared too late for Nernst's consideration) was from a lecture at the 14th Meeting of the Bunsen Society from May 9–12 in Hamburg (Nernst 1907). In it, he recapped his results from his heat theorem (Nernst 1906), including the discrepancy in the temperature between Haber's measurements and his own theoretical determination. At the equilibrium concentration of $p_{NH_3}=0.12\times10^{-3}$ atm (or 0.012 vol.-%), Nernst had calculated a temperature of 893 K while Haber had reported 1293 K (compare Figs. 11.7 and 11.8). According to Nernst, the ammonia yields reported by Haber in 1905 were unreliable because they had been measured at atmospheric pressure and were too small to provide accurate results.

In order to improve accuracy, Nernst and Friedrich Jost, citing the law of mass action, obtained higher yields using an electric pressure oven during the winter of 1906–1907. The higher pressure would also increase reaction rates and bring the system into equilibrium more quickly. They used a platinum wire catalyst and, like Haber and van Oordt, H_2SO_4 to extract and assess the amount of ammonia in equilibrium with nitrogen and hydrogen. In the publication, only ammonia formation experiments were reported, as Haber had previously noted after receiving Nernst's letter. Dissociation was not discussed. Despite lacking these complementary measurements, the independence of the results from gas flow rates convinced Nernst and Jost that they had measured the true equilibrium position.

Although Nernst's publication was meant as an overview for the conference, the presentation of his results is notable. In reporting on the observed values of the partial pressures of ammonia obtained at different temperatures, the usually precise researcher gave only the averages from many experiments. He did not elaborate on the composition of the precursor gas mixture or exact experimental pressures. A "precisely calibrated manometer"[26] was used to measure the pressure, which was between 50 and 70 atmospheres, and Nernst had to calculate the equilibrium constant for 1 atm. These were Nernst's first experimental results (Table 11.4)—up

---

[26] "...genau geeichter Manometer..."

Table 11.4 Values for ammonia equilibrium given by Walther Nernst in 1907 (Nernst 1907). t is the temperature in °C, T in K, $\sqrt{K}$ the square root of the equilibrium constant, and $100 \cdot x$ obs. and $100 \cdot x$ cal., 100 times the observed and calculated values for $x = p_{NH_3}$ ($100 \cdot x = $ Vol.-%). The calculations are based on Eq. 11.8. For further details see Jost (1908a) in Sect. 11.1.5

t [°C]	T [K]	$\sqrt{K}$	$100 \cdot x$ obs.	$100 \cdot x$ cal.
685	958	1830	0.0178	0.0196
809	1082	3783	0.0087	0.0082
836	1109	4460	0.0072	0.00702
876	1049	5900	0.0055	0.00561
920	1193	7560	0.0043	0.00448
1000	1273	10200	0.0032	0.00308
1040	1313	12170	0.0026	0.00261

until this point he had only published theoretical findings. However, the data and description were not specific enough to allow his experiments to be reproduced. Jost's 1908 publication of the same study provided the details missing in Nernst's report (Jost 1908a).[27] The expansion of published information meant that Nernst's overview played a similar role to Haber and van Oordt's 1905 preliminary letter, namely, establishing a presence in the field of ammonia synthesis from the elements.

Fitting the data points to thermodynamic theory, Nernst made use of Eq. A.27, which contained the simplified expression for the heat of formation of ammonia from Eq. A.25. He obtained:

$$log K = -\frac{Q}{4.571\,T} + B = -\frac{6130}{T} + 12.86 \tag{11.7}$$

or

$$log x = +\frac{3065}{T} - 6.918 \tag{11.8}$$

From this equation, he determined the heat of reaction for two moles of NH_3 to be 28,020 cal at room temperature. His results contradicted Thomsen and Berthelot's determination of 24,000 cal, which had been used by Haber in 1905 and 1907, and by Nernst himself in 1906. With this new value, combined with Haber and van Oordt's 1905 partial pressure of ammonia, $x = 0.12 \cdot 10^{-3}$ atm (0.012 Vol.-%), Eq. 11.8 indicated a temperature of 1023 K. According to Nernst, the result was "in far better agreement"[28] with his theoretical value of 893 K than Haber's original

---

[27] For comparison with Haber's values, full consideration of Nernst and Jost's work is given in Sect. 11.1.5 and Fig. 11.10 in the context of Jost's own publication (Jost 1908a). However, for the next section it is necessary to have some knowledge of these results.

[28] "...in weit besserer Uebereinstimmung..."

value of 1293 K (Nernst 1906).[29] Referring to his heat theorem, if he had chosen a value higher than 24,000 cal in his 1906 calculations, the temperature resulting from Eq. 11.4 would have been "practically in full agreement" with 1023 K.[30] However, the controversy on the equilibrium of ammonia was not yet settled.

The final two pages of Nernst's 1907 publication consist of an excerpt from the discussion following the presentation of his results at the Bunsen Society meeting. The exchange has taken on a central role in the history of ammonia synthesis to illustrate the dynamic between Walther Nernst and Fritz Haber. As indicated in Part I, there is more behind the circumstances of their interaction than only the vitriol and even subjugation suggested in an array of historical and popular literature. While the discussion was pointed and both men were defensive, it also facilitated the combination of different fields of knowledge: Haber, the application-oriented experimentalist, and Nernst, the fundamental truth-oriented theorist, engaged in an exchange of ideas. However, even this description oversimplifies the two men's expertise and intentions.

11.1.4 14th Meeting of the Bunsen Society, Hamburg, 1907

The discussion between Fritz Haber and Walther Nernst in Hamburg, published as the final two pages in Nernst's 1907 paper (Nernst 1907), has been cast as a dramatic showdown between the famous Prof. Nernst of Berlin and the little-known Prof. Haber of Karlsruhe (Coates 1937, p. 1652), (Goran 1967, pp. 46–50), (Suhling 1993), (Stoltzenberg 1994, pp. 154–156), (Szöllösi-Janze 1998b, pp. 167–169), (Charles 2005, pp. 87–91), (Bartel and Huebner 2007, p. 167), (Erisman et al. 2008; Hager 2008), (Sheppard 2020, chapter 5).[31] Nernst's direct criticisms and Haber's anxious persona are used to illustrate a situation in which a senior researcher publicly humiliated his junior colleague and in doing so, provided him with the sharpened physical insight that sent him on his way to a scientific breakthrough.

I will present the discussion in Hamburg in a different light by detailing the motivation and state of knowledge of both scientists.

The interaction at the Bunsen Society meeting is justly described as a crucial moment in the history of ammonia synthesis. The exchange of insight, the dynamic between experiment and theory (as discussed in Part I, Chap. 6), and Nernst's statements indeed provided the impetus for Haber to perform new measurements. However, Haber's comments in Hamburg were also of great relevance. After the

---

[29] Although it was not remarked until later, Haber's 1907 value of 0.0048 Vol-% at 1273 K and his "lower limit" nickel measurement from 1905 were in even better agreement with Nernst's 1907 value of 0.0031 Vol.-%.

[30] The actual calculated temperature in this case is 997 K.

[31] Although presenting a different viewpoint than my own, the narrative in Sheppard (2020) is substantiated with both archival evidence and physicochemical argumentation.

publication of his 1907 results, Haber was less influenced by Nernst's comments than existing literature leads us to believe. As we have seen, Haber's approach to ammonia synthesis had already undergone a transformation by this time, and he was now confident in his results. The discourse was rather an interaction between two men, both mature in their physicochemical understanding, who had grasped the profound meaning of the initial experimental and theoretical results on ammonia equilibrium.

This important aspect, absent from the literature, is integral in examining Haber and Nernst's discussion. They were the only two scientists in that time who had a detailed understanding of how the equilibrium of ammonia with nitrogen and hydrogen had been (approximately) determined. They also both understood that a successful industrial application depended on the behavior of the system at lower temperatures and on the identification of a more effective catalyst. The outcome could not yet be conclusively assessed: the inaccuracies in the thermal measurements used to extrapolate the data allowed for either conclusion. The examinations of this topic in the historical literature make it appear as if Nernst meant to discredit Haber's results because they were off by a factor of three or four. This deviation would have been significant if there had been an established expectation of what the equilibrium value should be. But there was not. The earlier numbers obtained by Ramsay and Young, Le Chatelier, Ostwald, and Perman were either inaccurate or widely fluctuating. Or Both. Furthermore, some of their results were not based on any actual ammonia production at all. It is clear from their discussions that none of these researchers expected the small amount of ammonia Haber measured. Nernst's critical stance was rather due to his limited data (his calculations and initial measurements) and he still had reason to doubt the accuracy of his own reported results.

It must be emphasized again that Haber's 1905 numbers established a reliable order of magnitude for what had previously been a completely unknown quantity. There was also still no experimental evidence that the ammonia system obeyed Le Chatelier's principle (König 1954). Before Nernst's calculations and Haber's measurements, the equilibrium of ammonia could have been, according to the expectations at the time, orders of magnitude away from what Haber had found. When Nernst completed the calculations for ammonia with his heat theorem and saw the similarity to Haber's results (even if off by a factor of three or four) he must have immediately understood what it meant. The results of their independent methods had suddenly converged with an accuracy that could not have been accidental. If Nernst had not been convinced, he would have had no reason to contact Haber. He would not have written a letter to the junior scientist if he thought Haber's result was nothing but another in a long line of erroneous measurements on ammonia. That is, he did not engage Haber to teach him a lesson, but rather he knew Haber's results were of meaningfully accuracy and perhaps just favorable enough for industrial application. Nernst wanted in on the action.

The published excerpt of the discussion begins with Haber covering his 1905 results with van Oordt in light of the presentation made by Nernst. His older study was "not carried out in order to determine the equilibrium of ammonia [but rather to determine industrial feasibility]." It was Nernst's letter "and the verification of [Nernst's] theory...that prompted [Haber]...to supplement the few, older measurements made with crude means with a large number of [new measurements] made with the utmost care..."[32] Haber took the same approach as in his 1907 paper (Haber and Rossignol 1907e) by describing a range of values and noting the 1907 numbers were closer to those measured by Nernst and Jost. At 1000 °C (1273 K) Haber reported the new value as 0.0048 Vol.-%, while Nernst measured 0.0031 Vol.-%. Haber also drew attention to the consequences of the uncertainty in the heat of formation of ammonia and its temperature dependence.

Nernst agreed that the outcomes of the new experiments were more acceptable but still cast doubt on the accuracy Haber was able to achieve with low yields at atmospheric pressure. "I would like to propose," Nernst responded, "that Herr Prof. Haber now use, instead of his earlier method...a method [under pressure] that will surely result in truly accurate values due to the higher yields."[33] This comment, that the use of pressure favored ammonia generation at equilibrium, is often cited as Nernst's lesson, or "decisive advancement in insight" (Szöllösi-Janze 1998b, p. 167),[34] to Haber. However, it was nothing more than a suggestion: Haber, like Wilhelm Ostwald before him,[35] understood the effects of working at higher pressures and had made explicit reference to it in 1905 (Haber and van Oordt 1905b, p. 344). By 1907, Haber was an established physicochemist. The expressions in Appendix A, taken from his book *Thermodynamik technischer Gasreaktionen* (Haber 1905b), show his mathematical and theoretical proficiency, including when describing the effect of pressure. The book had been well received and his publications had reached an international audience–in some ways Haber was even more a modern theorist than Nernst. He was comfortable with the concept of entropy and often used it in publications and correspondence,[36] which helped break down chemists' inhibitions toward its application (Coates 1937, p. 1650). Nernst, on the other hand, who decided not to use entropy to express his initial formulation of the third law, "thought the notion of entropy lacked concreteness and was inappropriate (Haberditzl 1960, p. 408), (Suhling 1972)."[37] In short, Haber's

---

[32] Eine Arbeit "welche nicht der Bestimmung des Gleichgewichtes wegen unternommen wurde [...] und die Prüfung seiner [Nernsts] Theorie...hat mich...veranlasst...die seiner Zeit mit gröberen Hilfsmitteln ausgeführten wenigen Versuche zu ergänzen durch eine grosse Anzahl derselben, die mit möglichster Sorgfält durchgeführt wurden..."

[33] "...ich [möchte] doch vorschlagen, dass der Herr Prof. Haber statt seiner früher angewandten Methode...doch nun auch eine Methode [unter Druck] anwendet, die wegen der grossen Ausbeute wirklich präzise Werte geben muss."

[34] "...entscheidender Erkenntnisfortschritt..." The formulation also implies Haber did not understand the effect of pressure before 1905/06. See also Suhling (1993, pp. 349–350).

[35] See page 84.

[36] See Haber (1905b, chapter 2) or Haber (1907c).

[37] "...hielt den Entropiebegriff für unanschaulich und unzweckmäßig...".

derivations contained the same mathematical rigor as Nernst's, and he was aware of the theoretical effect increased pressure would have on his ammonia experiments.[38]

It is rather Haber's reaction to Nernst's suggestion that gives us insight into why there was a deviation between their respective numbers, that is, why Nernst's measurements of ammonia equilibrium were lower than Haber's. "With Herr Geheimrat Nernst's apparatus," Haber said, "equilibrium can only be reached from one side, while I can reach it simultaneously from both sides."[39] Haber had mentioned this opinion in his 1907 publication and Nernst's response now in Hamburg was that he *could* reach equilibrium from both sides in his pressure oven–but not that he had done so. Haber's point was that during ammonia formation experiments from nitrogen and hydrogen, a sub-optimal catalyst (such as Nernst's platinum compared to Haber's iron) would not increase the reaction rates to those found in dissociation reactions because of a large energy barrier to formation. Nernst's measured yields would then be smaller than the true equilibrium value. This point remained a target of Haber's criticism and Nernst later changed his experimental approach.

At the end of the excerpt, Nernst reassessed the state of ammonia synthesis in reference to an industrial context in what has become a famous declaration. However, rather than only attacking Haber, he was also differentiating between Haber's 1905 and improved 1907 results.

> It is regrettable that the equilibrium of ammonia formation is shifted further toward the low end of the yield than one would have supposed from Haber's highly inaccurate numbers. We could have then considered synthetically producing ammonia from hydrogen and nitrogen. But the situation is actually much less advantageous, the yields are roughly three times smaller than expected.[40]

While Haber's skepticism toward the industrial potential of direct ammonia synthesis from the elements remained (he still preferred the electric arc into 1908 and 1909 during his initial work with BASF), Nernst's seemingly pessimistic attitude was not exactly as it seemed, for he was a business man. In 1897, while still in Göttingen, he had developed an incandescent lamp based on a mixture of zirconium and yttrium oxides that generated light more efficiently than the carbon filament lamp and did not require a vacuum. It became known as the Nernst-Lampe and the patent was sold to the AEG Aktiengesellschaft, providing Nernst with substantial financial gain. After a successful marketing campaign, it sold millions of units over a decade beginning in 1903 (Bartel 1989, pp. 45–49). Ammonia was simply another opportunity. In September of 1906, almost a full year before the meeting in Hamburg, Nernst had entered into a contractual agreement with the

---

[38] See, for example, Haber (1905b, p. 59, Eq. 3).

[39] "Man erreicht mit dem Apparate von Herrn Geheimrat Nernst das Gleichgewicht nur von der einen Seite, während ich es gleichzeitig von beiden Seiten erreiche."

[40] "Es ist sehr bedauerlich, dass das Gleichgewicht nach der Seite der viel geringeren Bildung mehr verschoben ist, als man nach den stark unrichtigen Zahlen Habers bisher angenommen hat, denn man hätte wirklich daran denken können, Ammoniak synthetisch herzustellen aus Wasserstoff und Stickstoff. Aber jetzt liegen die Verhältnisse sehr viel ungünstiger, die Ausbeuten sind ungefähr dreimal kleiner als zu erwarten war."

company Griesheim-Elektron to investigate the upscaling of ammonia synthesis from the elements (Farbwerke Hoechst 1966, pp. 19–24). Haber would not learn of this partnership until 1908 in Berlin, where Nernst himself informed Haber of his relationship with industry. Not surprisingly, this meeting led Haber to strengthen his own industrial pursuits. By that time, Nernst's experiments were underway and seemed promising, however, any success they may have had was not formulated into a patent before BASF's own patent application with Haber was submitted. In reaction, Nernst's work became part of an ultimately unsuccessful nullification suit (Suhling 1993), (Stoltzenberg 1994, pp. 170–180), (Szöllösi-Janze 1998b, pp. 181–185).

Nernst looked not only to better his position where he could financially, but was also known to accept praise for scientific endeavors, even if his actual role was unclear. When Le Rossignol met Nernst years later, he acknowledged Nernst's contribution to the work on ammonia synthesis: "Herr Geheimrat, it was also with your help,"[41] the engineer remarked. Although Nernst had admitted by that time that he could not have performed the experiments any better himself, he took the compliment and "swallowed it like butter (Jaenicke 1959)."

The evidence that Nernst was more optimistic about industrial ammonia synthesis than he let on is not only provided by the possibilities inherent in Haber's measurements. It is also found in Nernst's theoretical results. The chemical constant for hydrogen that Nernst used in his theory (the number 2.2 that appears multiplied by a factor of 3 in Eq. 11.3) was determined by the thermal measurements available to him at the time. But in 1909, he revised the value to 1.6 (Nernst 1909). Later, Haber remarked on the influence of the different values (Haber 1924, p. 16, reference 5 (pp. 22–23)):

> If one asks what the consequence would have been had [Nernst] at the time (1906–07) used the value of 1.6, which he later recognized to be true, in otherwise exactly the same conditions and methods of calculation, one can easily see he would have calculated an 8-fold increase in the ammonia yield. *Thus, he would have come to the conclusion that my estimate was not to high, but rather too low* [italics added] [. . .] These circumstances become more acute through the fact that at the time of Nernst's calculation neither the heat of formation of ammonia nor the specific heat were adequately known.[42]

Thermal data was again at the root of discrepancies between experiment and theory. The issue was not only Nernst's chemical constants used to calculate the integration constant, but also the heat of formation, Q. The value varied between 12,000 and 14,010 cal/mol,[43] and approximating its temperature dependence was a constant trade-off between manageability (complexity) and accuracy. The con-

---

[41] "Herr Geheimrat, es war auch mit Ihrer Hilfe."

[42] "Fragt man, was die Folge gewesen wäre, wenn [Nernst] seinerzeit (1906–07) alsbald den von ihm später als richtig erkannten Wert von 1,6 bei sonst völlig gleichen Voraussetzungen und gleicher Rechenweise benutzt hätte, so sieht man leicht, daß er einen 8 mal größeren Gleichgewichtsgehalt an Ammoniak errechnet und damit zu dem Schlusse gekommen wäre, daß meine Schätzung nicht zu hoch, sondern zu niedrig sei [. . .] Dieser Sachverhalt wird noch stärker durch den Umstand herausgehoben, daß zur Zeit der ersten Nernstschen Berechnung weder die Bildungswärme des Ammoniak noch seine spezifische Wärme ausreichend bekannt waren."

[43] The modern value is 10,990 cal/mol.

sequences for the value of Eqs. A.23 and A.27, for example, are noted in the next section. Remembering the attention to detail with which Nernst assessed his chemical constants in section two of his 1906 paper (Nernst 1906), it is difficult to accept that he did not consider the effect of deviations from the values he chose.[44] Nernst was aware of the uncertainty in thermal data that made a determination of the equilibrium yield at lower temperatures difficult. The equilibrium position and its temperature dependence, as it was speculated in 1907, did not exclude the possibility of industrial production—especially if one allowed for the error in thermal measurements. Nernst must have seen this in the numbers, otherwise he would not have approached Griesheim-Elektron.

Although Nernst's assertions at the Bunsen Society meeting may be viewed as a deterrent to Haber or to others who had come to similar conclusions about industrialization (Szöllösi-Janze 1998b, p. 175), they should not be viewed as an unusually acrimonious attack. Nernst was not opposed to polemics at the highest level and was known to criticize friends and colleagues if he felt it necessary (Bartel 1989, p. 50), (Szöllösi-Janze 1998b, p. 220). Haber had also been involved in such disputes. The debate with Edgar Perman between 1905 and 1907 is one example and Haber frequently mentioned others in his correspondence along with his anxious reactions (Haber 1900a, 1901a,b,c, 1902a). "Apparently," wrote Margit Szöllösi-Janze in another example (Szöllösi-Janze 1998b, p. 111),

> Haber's actions [at the 5th Annual Meeting of the German Electrochemical Society in Leipzig in 1898] had a provocative effect, at least on individual researchers...Years-old feuds were easing, the fallout of which had been publicized, and excessively occupied the temperamental, yet thin-skinned, and increasingly nervous Haber. He created scientific polemics [but was] held back and wisely advised by Richard Abegg who discretely moderated the attacks from his friend.[45]

This comment is not meant to convey Haber's actions and correspondence as mere gripes, attacks, and defensive posturing. As with his comparison of the state of physical chemistry in Germany and the United States in Chap. 10, Haber often gave meaningful commentary in his letters on the general state of science, documenting human interactions and the pursuit of publications. Some of the correspondences were quite gracious, even, for example, when it came to the rejection of one of his

---

[44] The agreement between Nernst's measurements and theory have been described as "coincidental (zufällig)" (Wendel 1962, p. 27), (Szöllösi-Janze 1998b, p. 167, citation 72). However, Nernst's attention to detail and suspicious attitude toward the accuracy of thermal constants makes it possible he suspected the correct number for hydrogen lay somewhere within a range of values. He may have chosen a value that supported his measurements or he may have believed 2.2 was the value best supported by available data. Either way, the term "coincidental" does not describe well Nernst's approach to research.

[45] "Offenbar wirkte Habers Auftreten [auf der 5. Jahrestagung der Deutschen Elektrochemischen Gesellschaft in Leipzig im Jahr 1898] zumindest auf einzelne Fachvertreter provozierend...[Es] entspannen sich jahrelange 'Fehden,' die publizierten Niederschlag fanden und den temperamentvollen, aber dünnhäutigen und zunehmend nervösen Haber über Gebühr beschäftigten. Er verfasste wissenschaftliche Polemiken, gebremst und klug beraten von Richard Abegg, der die Attacken seines Freundes besonnen mäßigte."

papers at the hands of Nernst in 1900. In that situation, Haber could only agree with Nernst's reasoning (Haber 1900b).

Keeping with this balanced perspective, we find what has been portrayed in literature as Haber's riled reaction to the encounter with Nernst to be not unusually adverse; when viewed in the wider context of Haber's career it was part of a broad range of emotions. In fact, the junior scientist had the opportunity to change and approve the excerpt of the discussion in Hamburg that was printed in the *Zeitschrift für Elektrochemie*. "I hope to leave on Monday," Haber wrote to Richard Abegg[46] in an undated letter before leaving for England to meet his friend at the Seventy-Seventh Meeting of the British Association for the Advancement of Science in Leicester, England between July 31 and August 7, 1907[47,48] (Murray 1908, p. 480), (Zott 2002, p. 382, Letter 152: Abegg to Ostwald, March 29, 1907),

> Let me know your exact itinerary, if possible in writing. I have written to [Heinrich] Danneel that in the revision of the discussion with Nernst I would like to have my best sentence stricken which begins "I consider it a significant result of Nernst's theory but I must say that..."
>
> In case you have the revision, please strike it.
>
> The quarrel with Nernst has risen to a form that makes this necessary.[49]

---

[46] The undated letter is only addressed to a "dear friend (lieber Freund)," however, the content shows it was meant for Richard Abegg. Both Abegg, to whom Haber often referred to as "dear friend," and Heinrich Danneel were at the time the editors of the *Zeitschrift für Elektrochemie*, where the discussion from the Bunsen Society meeting in Hamburg was published.

[47] This was three months after the Bunsen Society meeting (May 9–12, 1907) and two days before the publication of the resulting discussion on August 9, 1907. The undated letter must have been written just prior to the conference in England.

[48] Although this letter is undated, several factors speak for its having been written in the summer of 1907 in regards to the printing of the excerpt from the Bunsen Society meeting and not in December in regards to Haber's forthcoming article on ammonia synthesis (Haber and Rossignol 1908f) as stated, for example, in Stoltzenberg (1994, p. 156) and Zott (2002, p. 399). The reason they dated the letter to December may have been the following: the content is similar to another letter from Haber to Abegg from December 2, 1907 (Haber 1907b), which was clearly in reference to Haber's 1908 publication. In addition, the letter is found at the end of the file HS 848 in the Archive of the Max Planck Society among other letters from December. However, the trip to England was earlier in the year. In a letter to Richard Abegg from July 25, 1907, Clara Haber wrote she was not able to come to England with Haber where he had been invited to give a talk on gas reactions at the same meeting where Haber planned to meet Abegg (Haber 1907a), (von Leitner 1993, pp. 115–116). Furthermore, in the undated letter, Haber referred to the "discussion" with Nernst–there was no such discussion in his 1908 publication–and the sentence which Haber wanted removed used the first person singular. In his 1908 publication Haber continuously used "we" to refer to both Le Rossignol and himself.

[49] "Ich hoffe Montag Abend hier fortzukommen. Teile mir genau deine Reiseroute und Zeit mit, eventuell schriftlich. Ich habe [Heinrich] Danneel geschrieben, daß ich an der Revision der Discussion mit Nernst über NH$_3$ meinen besten Satz in der Revision gestrichen wünsche, welcher lautet 'ich betrachte es als einen wesentlichen Erfolg der Nernst'schen Theorie, aber ich muß sagen, daß...' Wenn du die Revision hast, bitte streiche es. Der Krach mit Nernst hat Formen angenommen, die das erwünscht machen."

The public humiliation described in the literature to which Haber was subjected could have been countered by Haber himself, had he felt it necessary. This was neither the first nor last time that pointed language was used or that Haber was alerted to the content of a forthcoming publication on ammonia synthesis (results and commentary were often shared before printing) (Haber and Rossignol 1908g). Rather, his reaction was much the same as it had been to the other polemics he had had over his career: they caused him to vent in both public and private, and to redouble his efforts in the laboratory. The latter was not without reason. After the Bunsen Society meeting, there was no longer any doubt whether the equilibrium position of ammonia had been measured (or calculated)—only the accuracy of the measurements were still in question along with experimental details. And, of course, whether the process was suitable for industrial application. As for theory, Nernst had provided the missing concept that solved a central problem in chemical thermodynamics: the constant of integration in the formula for the free energies could now be calculated. But his results were only legitimate if backed by experiment, and the ammonia system had not conformed. After Hamburg, though, the deviations were no longer questioned on physical grounds; it was only the uncertainties in thermal data which posed a problem (Rossignol 1928).

Experimental work continued.

11.1.5   Über das Ammoniakgleichgewicht (On Ammonia Equilibrium) by Friedrich Jost, 1908

In March of 1908, Friedrich Jost published his detailed results underlying Nernst's 1907 publication (Jost 1908a; Nernst 1907). As had been the case with Nernst in 1907, he began with a critique of Haber's 1905 results. He argued that it was the small yields in Haber and van Oordt's experiments that led to the fluctuating and inaccurate numbers. Jost sought to avoid this issue by using higher pressures in the pressure oven (the same as described by Nernst in 1907) and also corrected for oxygen in his precursor gases. Here Jost first explicitly stated he had determined the equilibrium from both sides–a shortcoming to which Haber had already thrice called attention.[50] It becomes clear here that the dissociation experiments were not included in those discussed by Nernst in Hamburg, but were added later. The dissociation experiments were fewer in number and, as is apparent from the summary of results, made under different experimental conditions than the generation experiments.[51] The initial study, reported in 1907, was made only with

---

[50] A report by Karl Jellinek from 1911 on ammonia synthesis under pressure also contained results only from ammonia generation reactions (Jellinek 1911). In his 1966 retrospective, Paul Krassa emphasized this difference in Haber's and Nernst's experimental approach (Krassa 1966).

[51] The table containing the dissociation results is also labeled incorrectly. It, too, reports ammonia generation.

a platinum catalyst and complementary experiments were later performed with iron and manganese. None of the dissociation experiments in the 1908 paper were performed with platinum. It is not clear when iron and manganese were first used, but it was likely after the meeting in Hamburg as a reaction to Haber and Le Rossignol's criticism. Jost's dissociation results do, however, support the trend seen in his generation experiments.

Jost summarized his large number of experiments made between 14.3 and 74.5 atm,[52] 876 and 1040 °C, and differing nitrogen-hydrogen mixtures into tabulated results, recalculated for 1 atm and a gas mixture of $N_2:H_2 = 1:3$. These are given in Table 11.5 and Fig. 11.10 as black squares. Jost fitted the same two-term equation as Nernst in 1907 (Eq. 11.8), which is shown in Fig. 11.10 as a solid line.

Jost also interpreted his results in terms of Nernst's heat theorem. In 1906, Nernst did not yet have evidence the heat of formation, Q, of ammonia was higher than 12,000 cal/mol at room temperature, so he used exactly that value. However, Jost's data was best fit with the assumption that Q was 14,000 cal/mol at working temperatures, which implied a somewhat higher value than 12000 cal/mol at room temperature. Jost also used the approximation for Q in Eq. A.22 while Nernst had used the simplified expression from Eq. A.25. Although Jost's approximation for Q did not completely match experimental data—it was lower than the measured value at working temperatures (1150 K) but higher than the "classic" value of 12,000 cal/mol at room temperature—it had "adequate accuracy."[53] The chemical

Table 11.5 Values for ammonia equilibrium given by Friedrich Jost in 1908 (Jost 1908a) (black squares in Fig. 11.10). t is the temperature in °C, T in K, $\sqrt{K}$ the square root of the equilibrium constant, and $100 \cdot x$ obs. and $100 \cdot x$ cal., 100 times the observed and calculated values in Eq. 11.8 for $x = p_{NH_3}$ ($100 \cdot x$ = Vol.-%), respectively

t [°C]	T [K]	$\sqrt{K}$	$100 \cdot x$ obs.	$100 \cdot x$ cal.	Catalyst
685	958	1830	0.0178	0.0196	Mn
705[a]	978	2160	0.0150	0.0165	Mn
790[a]	1063	3465	0.00937	0.01923	Mn
809	1082	3783	0.0087	0.0082	Mn
836	1109	4460	0.0072	0.00702	Mn
845[a]	1118	4890	0.00664	0.00665	Fe
876	1049	5900	0.0055	0.00561	Pt
920	1193	7560	0.0043	0.00448	Pt
976[a]	1249	9145	0.00355	0.00344	Fe
1000	1273	10200	0.0032	0.00308	Pt
1040	1313	12170	0.0026	0.00261	Pt

[a]Experiments in which NH_3 was dissociated, the rest are generation experiments

[52] The lower pressure experiments (14–15 atm) were meant to substantiate use of the ideal gas approximation valid at higher pressures–that is, he showed low and high pressure experiments gave compatible results.

[53] "...hinreichende Genauigkeit..."

constant for ammonia formation was now 6.62 instead of 6.6 in Eq. 11.3; the constant for hydrogen remained 2.2. Nernst's 1906 equation for the equilibrium constant, Eq. 11.4, became:

$$\log K = log \frac{0.325}{x} = -\frac{2571}{T} + 1.75 \, log \, T - 0.000385 \, T + 1.29 \qquad (11.9)$$

and is shown in Fig. 11.10 by the small-dashed line. Jost compared Eq. 11.9 with the experimental results in Table 11.5 and was satisfied with the agreement.[54] "The calculation," he wrote with foresight,

> is tentative until we have better knowledge of the specific heat of ammonia and of the "chemical constants" [in Eqs. 11.3 and 11.9] for nitrogen which are. . .the most uncertain for this gas[55]

Jost's expression for Q indicated a room temperature value of 12,753 cal/mol for the heat of formation of ammonia—higher than the value of 12,000 cal/mol accepted at the time. He also noted that if Q_o equaled 12,753 cal/mol in Eq. 11.4, then Nernst's 1906 results would have been in better accordance with experiment (the dashed-dotted line in Fig. 11.10) because a higher value of Q_o led to a higher calculated ammonia yield. Comparing his results to Haber and Le Rossingol's 1907 numbers, Jost still found inconsistencies. "I cannot," he wrote, "give an exact single cause for the considerable deviation. . .however, I find no reason to view my results. . .as unreliable."[56]

There were conspicuous differences between this paper from 1908 and Nernst's 1907 conference contribution. The changes went beyond the application of Nernst's heat theorem; they were also experimental in nature. The importance of measuring the ammonia system "from both sides" and the use of different catalysts had become clear to Jost and Nernst after the Bunsen Society meeting (but Nernst never considered the catalyst to be as important to understanding ammonia synthesis as Haber did (Mittasch 1951, p. 69)). Nevertheless, through their interaction with Haber, Nernst and Jost learned more of the reality of meticulous experimental work.

This and later publications show that Haber was not the only one who had learned something in Hamburg.

---

[54] This statement was corrected in Jost (1908b). Originally he reported his *dissatisfaction*.

[55] "Die Rechnung is so lange eine provisorische, bis wir nähere Kenntnis über die spezifische Wärme von Ammoniak, ebenso der "chemischen Konstanten" [in Gleichungen 11.3 und 11.9] für Stickstoff, die gerade für dieses Gas. . .am unsichersten sind, haben."

[56] "Eine bestimmte Einzelursache für die beträchtlichen Abweichungen. . .anzugeben, ist mir nicht möglich. Auch finde ich keinen Grund, meine Zahlen. . .für unzuverlässig zu halten."

11.1.6   Bestimmung des Ammoniakgleichgewichts unter Druck (Determination of Ammonia Equilibrium Under Pressure) by Fritz Haber and Robert Le Rossignol, 1908

During the second half of 1907, Haber and Le Rossignol had worked "feverishly" on the third round (second round with Le Rossignol) of experiments to assess the equilibrium position of ammonia, this time under pressure as Nernst had suggested (von Leitner 1993, p. 125). The new apparatus was different from Nernst's and depended significantly on Le Rossignol's engineering abilities. For example, he built the needle valves from scratch that were needed for the accurate dosing of gases (Krassa 1966; Travis 1993a). As Le Rossignol recollected years later, "Haber was not good as experimentalist [sic]...[he] did the theoretical side, and I did the engineering side (Jaenicke 1959)."

They were not working out of a need to appease Nernst. Haber was convinced his numbers were correct, and he knew why Nernst's were not. They published in April of 1908 (Haber and Rossignol 1908f), beginning with a blunt assessment. "The question," wrote Haber, "of the exact equilibrium position for the reaction $N_2 + 3\,H_2 \leftrightarrows 2\,NH_3$ is, between Herr Nernst and ourselves, controversial."[57] He was also candid in regards to the quality of the measurements. "The result of these determinations," Haber continued,

> ...shows that Nernst's objection is not justified; the values determined [in our current study] under [30 atm] pressure completely confirm our earlier conclusions. The results of the experiments carried out and reported last year by Herr Nernst and his pupils [Jellinek and Jost] deviate from our findings to a degree that lies outside the margin of error. The values communicated by Herr Nernst have all been determined from the nitrogen-hydrogen side and lie below our values. We must conclude, therefore, that equilibrium was in no way reached during these experiments. The numbers communicated by Herr Nernst have a different trend than ours: they decrease faster with increasing temperature.[58]

It was the same opinion Haber had expressed in Hamburg: Nernst had not properly determined the equilibrium position because he had only performed generation experiments (with a platinum catalyst). Furthermore, after having constructed their own pressure oven, Haber and Le Rossignol suspected Nernst's apparatus had not given accurate temperature readings at the point of ammonia production. Not only

---

[57] "Zwischen Herrn Nernst und uns ist die Frage strittig, welches die genaue Lage des Gleichgewichtes der Reaktion $N_2 + 3\,H_2 \leftrightarrows 2\,NH_3$ ist."

[58] "Das Ergebnis dieser Bestimmungen...lehrt, daß der Nerntsche Einwand nicht berechtigt ist; denn die [in unserer Studie] unter [30 atm] Druck ermittelten Zahlen bestätigen vollkommen unsere früheren Resultate. Die Versuche, welche Herr Nernst mit seinen Schülern [Jellinek und Jost] ausgeführt und im vorigen Jahre mitgeteilt hat, weichen in ihrem Ergebnis von unserem Befunde in einem Maße ab, welches weit über die Fehlergrenze unserer Bestimmungen hinausgeht. Die von Herrn Nernst mitgeteilten Werte sind alle von der Stickstoff-Wasserstoff-seite [sic] her ermittelt und liegen alle tiefer als unsere Zahlen, so daß wir schließen müssen, daß bei diesen Bestimmungen das Gleichgewicht in keinem Fall erreicht worden ist. Die von Herrn Nernst mitgeteilten Zahlen besitzen einen anderen Gang als die unseren: sie fallen rascher mit steigender Temperatur."

were the equilibrium positions different, there was also a deviation in the overall dependence of equilibrium with temperature, meaning it was not simply a systematic error. As for the catalyst, Le Rossignol had repeated previous experiments with platinum instead of iron and found little catalytic effect, strengthening the argument that Nernst's generation experiments did not reflect equilibrium conditions (Krassa 1966). "We found out," said Le Rossignol later, "that Nernst's equilibrium was very bad (Jaenicke 1959)." Even porcelain shards (probably containing small amounts of iron) had a better catalytic effect than platinum (Haber and Rossignol 1908g). In a detailed account of erroneous sources of ammonia production, Haber and Le Rossignol also concluded that the commercially available nitrogen-hydrogen mixture used by Nernst and Jost contained a small amount of oxygen which turned to water and absorbed some of the ammonia during the experiment, possibly reducing the yield. After drying this commercially available gas mixture, they obtained the same results as with their own nitrogen and hydrogen from pre-dissociated ammonia.

Haber and Le Rossignol's new experiments were performed at different temperatures and pressures (22.35–30 atm) with iron and manganese catalysts, and an approximately 1:3 mixture of nitrogen-to-hydrogen. As in 1907, the increased reproducibility in comparison with the 1905 study is apparent. Average values for the experiments, as given in 1908, are collected in Table 11.6 and shown in Fig. 11.11.

The results were used to determine the constant in Eq. A.27, which differed from 1907 (Eq. 11.5) in a notable way: the heat of formation was set at 12,800 cal/mol:

$$log K_p = \frac{12800}{4.571\,T} - 6.06 \tag{11.10}$$

However, because this expression did not adequately represent all of the data from the previous study in 1907, Haber also considered an equation with an expression for Q from his 1905 book (Haber 1905b, pp. 202–205):

$$log K_p = \frac{2215}{T} - 3.626\,log\,T + 3.07 \cdot 10^{-4}\,T + 2.9 \cdot 10^{-7}\,T^2 + 4.82 \tag{11.11}$$

Table 11.6 Average values for ammonia equilibrium given by Haber and Le Rossignol in 1908 with a 1:3 N_2:H_2 mixture and iron and manganese catalysts (Haber and Rossignol 1908f) (black squares in Fig. 11.11). At 974 °C minimum values are given due to high gas fluxes; the numbers in parentheses denote the consideration of higher values found at lower fluxes

T [°C]	Vol.-%, 30 atm (measured)	Vol.-%, 1 atm (calculated)	$K_p \cdot 10^4$	Catalyst
700	0.654	0.0221	6.80	Fe, Mn
801	0.344	0.0116	3.56	Mn
901	0.207	0.00692	2.13	Fe, Mn
974	>0.144 (0.152)	>0.0048 (0.0051)	>1.48 (1.56)	Fe

where the heat of formation of ammonia at room temperature was again set at12,000 cal/mol. These expressions are found as the solid line (Eq. 11.10) and the dashed line (Eq. 11.11) in Fig. 11.11. "With this study," wrote Haber, "we hope to a certain degree to have brought the task of determining the equilibrium of ammonia, begun four years ago by Haber and van Oordt, to a conclusion."[59]

However, he still felt the need to rationalize the misinterpretation of his 1905 data and the equilibrium value that had been too high. Haber had done the same at the Bunsen Society meeting where Nernst was set on considering both Haber's 1905 and 1907 measurements instead of, as Haber would have preferred, only the 1907 results. It was part of Haber's trend of defending himself against any attack, even after the third set of experiments at high pressures had confirmed his earlier results. He was still wary of Nernst. In December of 1907, he sent Richard Abegg a copy of the manuscript of the 1908 paper and wrote, after clarifying the results (Haber 1907b):

> Nernst's values are, therefore, shown to be incorrect. Because in this publication I have to counter Nernst, whose sensitivity is known, I ask you as an unbiased expert and friend to read the...manuscript and alert me to parts which seem in any way harsh so that I can change them.[60]

This passage is at the end of the letter following scientific commentary on one of Haber's favorite topics: dilute solutions. Haber did not go into scientific detail on the subject of ammonia and instead restricted his comments to the political developments with Nernst. At this point in his investigations, his interest in ammonia synthesis (if only with regards to industrialization) had grown since 1903, and the lack of correspondence on the subject may be due to the confidence in his 1907 and forthcoming 1908 results. A lengthy quote from the end of the 1908 paper illustrates this attitude and also summarizes the narrative up to this point. "With regards to the fact," wrote Haber,

> that [before the first measurements] there was absolutely no knowledge about the equi-
> librium position [of ammonia], one will recognize that the order of magnitude of the
> values [calculated with Eq. 11.2] is correct,[61] and that the numbers were only used to
> justify two conclusions which are important and undisputedly correct. [These are] namely,
> "that from the beginning of red-heat onward no catalyst can produce more than traces
> of ammonia with the most favorable gas mixture if one works at normal pressure", and
> "even with very high pressure the equilibrium position still remains unfavorable" [...]
> The accuracy of the determinations of the position of equilibrium has gained a whole
> new importance through the theoretical developments of Herr Nernst. Uncertainties in the

---

[59] "Wir hoffen mit dieser Untersuchung die vor vier Jahren durch Haber und van Oordt in Angriff genommene Aufgabe, das Ammoniakgleichgewicht zu bestimmen, zu einem gewissen Abschluß gebracht zu haben."

[60] "Die Nernst'schen Werte erwiesen sich damit als unrichtig. Da ich in dieser Abhandlung Nernst entgegentreten muss, dessen Empfindlichkeit bekannt ist, so bitte ich Dich das...Manuskript zunächst als unparteiischer Fachgenosse und als Freund durchzulesen und mich auf Stellen aufmerksam zu machen, welche irgendwie scharf erscheinen, damit ich dieselben ändere."

[61] A footnote at this point in the original text reported the miscalculated constant of integration in Eq. 11.2 from 1905 (see page 111).

[initial] experiments. . .seem now to be unacceptable [. . .] [We will] in no way deny that the remaining difference [between Haber's measurements and Nernst's heat theorem] can be reduced through new investigations of the heat data which may result in a change in the calculated values according to Nernst's work. But we must absolutely counter the attempt to attribute the remaining difference to a deficiency in our experimental determinations as well as the underlying reason [which was] the experimentally oriented statement by Herr Nernst [to measure] the position of equilibrium [under pressure].[62]

Nernst's suggestion at the Bunsen Society meeting had not been necessary, though that only became clear after the fact. Fritz Haber's commentary, on the other hand, proved pertinent. He identified the reason for Nernst's lower experimental values—the measurements had only been made during ammonia production experiments—and offered a reason why Nernst's calculations nevertheless supported those experiments. It was due to uncertainties in thermal data. While both men still agreed the equilibrium position of ammonia was unfavorable (even at high pressures), it was only in public. Nernst had long since begun his relationship with Griesheim-Elektron. According to Le Rossignol's later account, it was the 1908 publication that prompted Haber and him to design an apparatus to continuously produce ammonia. The extrapolation of their data indicated an 8% ammonia yield at $\sim 600°$ and, although it pushed technological limits, a pressure of ~ 200 atm (Rossignol 1928), (Coates 1937, p. 1652).

In July, Jost published a letter responding to Haber and Le Rossignol's experiments under pressure (Jost 1908b). There were points of methodological criticism, including the assertion that Haber and Le Rossignol's assessment of the temperature was inaccurate. Mainly, though, it was a discussion of the discrepancies between their results, especially at higher temperatures. Jost did his best to draw them into agreement while noting, "there can be no discussion of a difference between Herr Haber and Herr Nernst, but rather only of such [a difference] between Herr Haber

---

[62] ". . .im Hinblick darauf, daß [vor den ersten Messungen] überhaupt keinerlei Kenntnis über die Gleichgewichtslage [von Ammoniak] bestand, wird man würdigen, daß die Größenordnung der [mit Gleichung 11.2] berechneten Werte richtig ist, und daß die Zahlen nur zur Begründung zweier Schlüsse benutzt wurden, die. . .wichtig und unbestritten richtig sind, nämlich 'daß von beginnender Rotglut aufwärts kein Katalysator mehr als Spuren von Ammoniak bei der günstigsten Gasmischung erzeugen kann, wenn man bei gewöhnlichem Drucke arbeitet', und 'auch bei stark erhöhtem Drucke bleibt die Lage des Gleichgewichtes stets eine sehr ungünstige' [. . .] Durch die theoretischen Entwicklungen von Herrn Nernst gewann die Präzionsbestimmung der Gleichgewichtslage eine ganz andere Wichtigkeit. Unsicherheiten der [ersten] Versuche. . .erschienen nun unerträglich [. . .] [Wir wollen] keineswegs die Möglichkeit in Abrede stellen, daß die verbleibende Differenz [zwischen Habers Messungen und Nernsts Wärmetheorie] durch erneute Untersuchung der Wärmegrößen, welche eventuell nach den Nerntschen Entwicklungen eine Aenderung des berechneten Wertes ergeben würde, kleiner herauskommen kann. Aber dem Versuche, die verbleibende Differenz einem Mangel unserer experimentellen Bestimmungen zuzuschreiben, und den zur Stütze dieses Versuches von Herrn Nernst geltend gemachten experimentellen Angaben über die Gleichgewichtslage [Messungen unter Druck durchzuführen], müssen wir durchaus widersprechen."

and me."[63] Whether Jost was trying to protect Nernst from criticism or whether he wished to distinguish himself from the senior scientist is not clear, especially considering Nernst himself had not pulled back from the debate. Either way, Jost's breadth of knowledge on the subject of ammonia synthesis is evident in his two publications from 1908 and shows that he was an independent researcher. At the end of the article, Jost suggested additional high temperature measurements as a solution to the remaining disagreement, but Haber and Le Rossignol already had something else in mind. An experimental and theoretical consensus was no longer the only objective.

---

[63] "Jedenfalls kann also nicht von einer Differenz zwischen Herrn Haber und Herrn Nernst, sondern nur von einer solchen zwischen Herrn Haber und mir gesprochen werden."

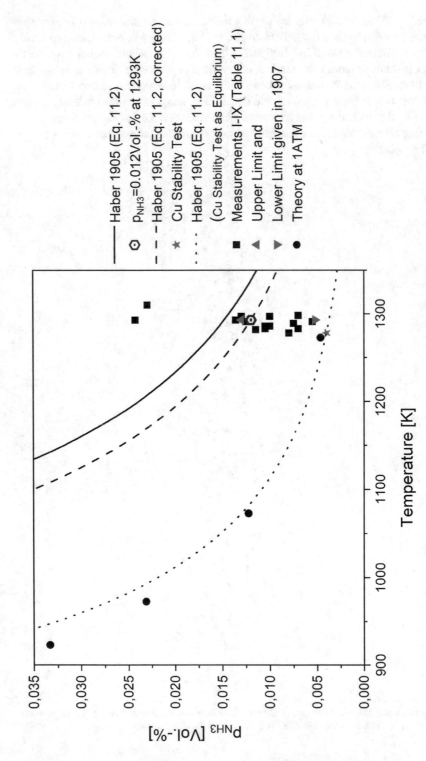

Fig. 11.7 Fritz Haber and Gabriel van Oordt's 1905 results

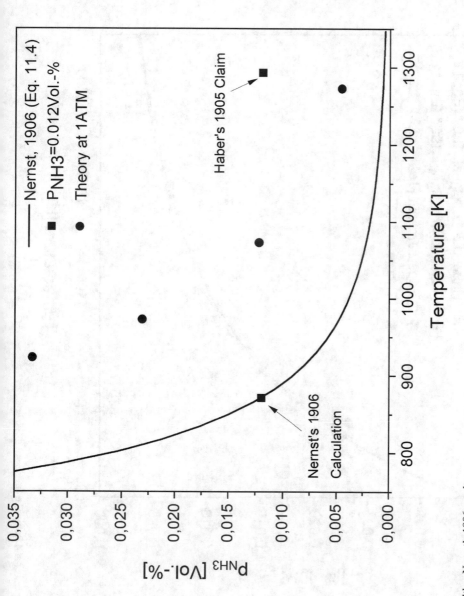

Fig. 11.8 Walther Nernst's 1906 results

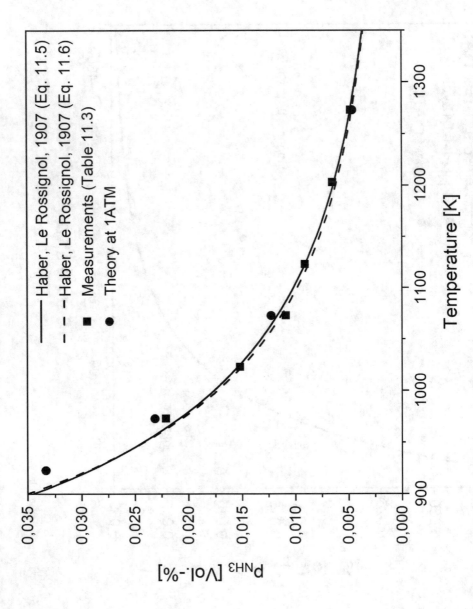

Fig. 11.9 Fritz Haber and Robert Le Rossignol's 1907 results

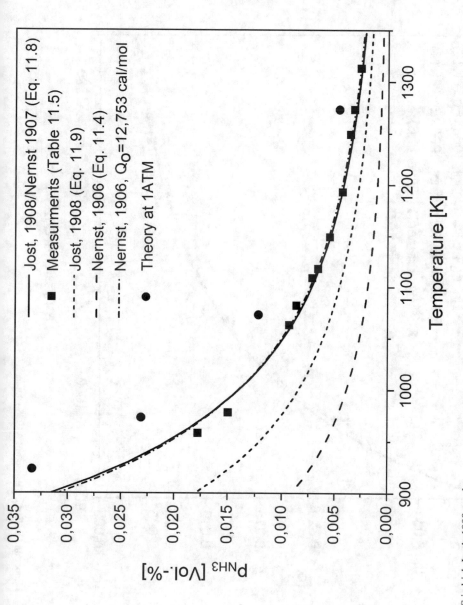

Fig. 11.10 Friedrich Jost's 1908 results

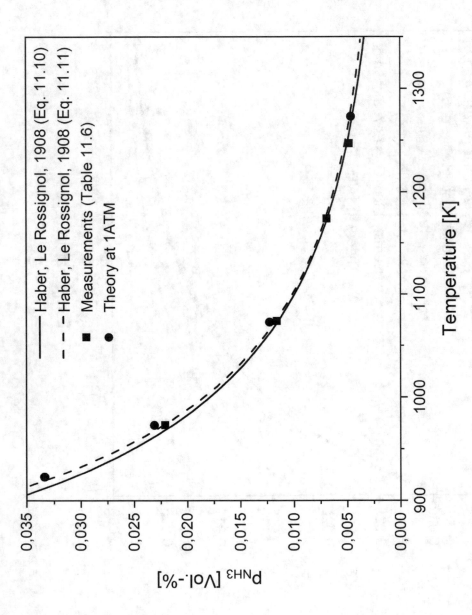

Fig. 11.11 Fritz Haber and Robert Le Rossignol's 1908 results

Chapter 12
Haber's Cooperation with BASF

Parallel to the experiments for the 1908 publication on ammonia synthesis, Haber continued to work on the electric arc with Adolf König (Haber and König 1907d, 1908e). At the time, BASF was stagnant; its ability to innovate had become too dependent on external academic research (Stoltzenberg 1994, pp. 146 147), (Szöllösi-Janze 1998b, p. 171), (Abelshauser et al. 2004, pp. 142–146). In December of 1907, Carl Engler, a BASF Board member, brought the company into contact with Haber in an attempt to remedy the problem. In January 1908, Haber wrote to Engler, mainly to recount his experiments with König (he and others still considered the electric arc to be the most promising method of fixing nitrogen) (Haber 1908b):

> I would like to respectfully point out that the fundamental factor moving me toward a relationship with industry is the need for apparatus and machines that are necessary for research on the problem and cannot be made available from public funds alone.[1]

It was not only the electric arc experiments that Haber felt were ready for cooperation with industry. He and Le Rossignol, based on their latest experiments, now wanted to build an apparatus for the continuous production of ammonia from the elements which could be upscaled (Krassa 1966; Rossignol 1928). Haber had been reassured ammonia synthesis at high pressures was "a promising undertaking"[2] not only by his work with Le Rossignol, but also by two techniques that had recently come to his attention: the liquefaction of air at 200 atm as well as manufacturing methods in the formate industry (Haber 1908a), (Haber 1920, p. 337). An academic setting would no longer be adequate for research on either the electric arc or direct ammonia synthesis; further development would have to take place in tandem with industry (Jaenicke 1958b).

---

[1] "...ich [möchte] noch ergebenst darauf hinweisen, dass das wesentliche Moment, welches mich zu einer Anknüpfung mit der Industrie drängt, das Bedürfnis nach Apparaten und Maschinen ist, welche für die Erforschung des Problems benötigt werden, und aus den staatlichen Fonds allein nicht angeschafft werden können."

[2] "...ein hoffnungsvolles Unternehmen..."

© The Author(s) 2022
B. Johnson, *Making Ammonia*, https://doi.org/10.1007/978-3-030-85532-1_12

Fig. 12.1 Fritz Haber in conversation with Heinrich von Brunck of BASF in 1908. By David Vandermeulen from *Fritz Haber: Les Héros* (Vandermeulen 2007); ©David Vandermeulen/Guy Delcourt Productions

The cooperation between Haber and BASF and the industrial upscaling of ammonia synthesis from the elements are well documented and need not be repeated here (Schwarte 1920, chapter 4), (Mittasch 1951, pp. 90–142), (Holdermann 1960, pp. 65–121), (Haber 1971, chapter 4), (von Nagel 1991, pp. 16–50), (Reinhardt 1993), (Stoltzenberg 1994, chapter 5), (Szöllösi-Janze 1998b, chapter 4), (Abelshauser et al. 2004, pp. 142–146), (Sheppard 2020, chapters 7 and 8). However, several points are pertinent to the scientific perspective (Fig. 12.1).

Haber and BASF entered into two contractual agreements on March 6, 1908 (Fig. 12.2). One was with Haber and König for the development of the electric arc process and the other was with Haber for the development of direct ammonia synthesis from the elements. BASF was still investigating the cyanamide process and it was only at the end of 1908 that their attention turned from this process and the electric arc to Haber's ammonia synthesis (BASF 1908b,c), (Mittasch 1951, p. 90). Throughout the first half of 1909, work continued on the design and construction of an apparatus that was not only capable of continuously producing ammonia at high pressures, but also exploited a heat exchange system. A wide array of catalysts, including iron, nickel, chromium, manganese, uranium, and osmium was also tested. On July 2, 1909, Haber and Le Rossignol were ready to present the result and

Fig. 12.2 Haber and Le Rossignol's laboratory apparatus for the continuous production of ammonia. *Source:* Archive of the Max Planck Society, Berlin-Dahlem, Jaenicke Sammlung, Picture Number XII.2/7

invited Carl Bosch, Alwin Mittasch and Julius Kranz (chief mechanic at BASF) to Karlsruhe for the demonstration (Fig. 12.2). It turned out to be a dramatic finish. A problem with a bolt during the initial assembly meant that the apparatus could not be put under pressure. The repairs took so long that Bosch eventually left; he did not see the historical presentation that followed. But Mittasch and Kranz were there to witness it and after the necessary repairs had been made the machine was turned on. At first nothing appeared to happen–but then the drops of liquid began to appear. Haber took a sample and demonstrated to the representatives from BASF: it was ammonia. He and Le Rossignol used uranium and osmium[3] catalysts

---

[3] Osmium had unexpectedly been identified as an effective catalyst several months before and BASF had considered purchasing from the Auergesellschaft its entire stock. At the time it amounted to nearly the world's entire supply of the element (Mittasch 1951, pp. 75–76), (Jaenicke 1958a,b), (Stoltzenberg 1994, pp. 158–160), (Szöllösi-Janze 1998b, pp. 177–179).

and produced ammonia continuously for several hours between 600–900°C at a pressure of ~175 atm (Haber 1909b), (Mittasch 1951, p. 77), (Sheppard 2020, chapter 8). It was here that all the elements in Ostwald's "recipe" from his 1900 patent application, established on the basis of physical chemistry, were brought together in one working system. It was a technological success and formed the basis for BASF's prosperity in the nitrogen market.[4]

Money would also complicate the arrangement as it had done in 1904 (Reinhardt 1993, 299–301), (Stoltzenberg 1994, pp. 144–151, 179–180), (Szöllösi-Janze 1998b, pp. 172–175, 180, 189–191, 481–487). Through a series of contracts and agreements, Haber and BASF arranged funding for equipment and compensation for his assistants and for Haber himself. For nitrate production, BASF agreed to pay 10,000 marks after the Haber-König patent was issued and 6,000 marks per year for further work. In addition, the company would finance equipment for three years and 10% of net profits would go to the inventors for the life of the patent. The agreement was soon renegotiated and by the end of 1908, Haber agreed to 23,000 marks per year for 10 years for ongoing research. For work on direct synthesis from the elements, Haber received a flat sum of 8000 marks in addition to the 10% stake. Haber made a further agreement with Le Rossignol that the engineer would receive 40% of profits from any future technical applications. By the time Haber and Le Rossignol demonstrated their apparatus in Karlsruhe in 1909, the ammonia "account" was 4400 marks overdrawn, and Haber requested an additional 8000 for catalyst research. BASF granted the entire sum. Due to substantial changes in the preferred method for fixed nitrogen production as well Haber's appointment as director to the Kaiser Wilhelm Institute for Physical Chemistry and Electrochemistry, a new agreement went into effect on November 1, 1911 in which Haber would receive 10,000 Marks per year for his services.

When ammonia production began at Oppau in 1913, the 10% stake in production became the subject of debate. While Haber understood it to be 10% of net profits, BASF insisted that it meant 10% of the difference between their profits from the Haber-Bosch-Process and what they would have earned had they continued other methods of ammonia production (the multi-step process). In consideration of production costs and sales prices at the time, this meant Haber would receive only two pfennigs per kilogram ammonia instead of five. He claimed that BASF had not properly informed him of this disparity during contract negotiations, but ultimately he was able to reach a compromise with the company. He would receive a flat sum of 1 to 1.5 cents per kilogram ammonia, depending whether it was produced directly by BASF or under license, with which he hoped to secure a growing percentage as production prices dropped. In 1919, the rate was lowered to 0.8 pfennigs per kilogram, with compensation to come at a later date (post-war circumstances caused Haber to fear accusations of war crimes which could result in the seizure of his assets (see Part I, Chap. 8)). Payments continued into the 1920s and amounted to several million marks. It was only ten years later, during the period of hyperinflation, that

---

[4] See Part I, Chap. 7 for the contributions of Carl Bosch and Alwin Mittasch to this outcome.

the agreement was no longer advantageous to Haber. In the end, an arrangement was reached in 1923 in which Haber likely received a one-time compensation in addition to the per-kilo share (now worth essentially nothing). The size of the sum is not known, but it seems to have been appropriate. In addition, Haber was offered a position on BASF's board of directors for which he received remuneration.

Diverging economic and academic aspirations also affected the dynamic between Haber and BASF when it came to the dissemination of scientific knowledge. The contractual negotiations between Haber and BASF included the publication of Haber's laboratory results and future disclosure of progress made while working with BASF. Haber had requested that any such arrangement be left out of the contract and that he and BASF discuss publication on a case-by-case basis; Haber felt he needed to accept some form of agreement in this respect, although it went against his conviction as a scientist (Engler et al. 1909; Haber 1909a). BASF agreed to Haber's initial request because the paper, according to an internal memo from August Bernthsen, a member of the board of directors at BASF, was "mainly of a polemic nature…that should serve to demonstrate that [Haber] is correct in the face of the attacks by Professor Nernst."[5] Haber acted quickly to publish in 1908 (BASF 1908a; Haber 1908c; Haber and Rossignol 1908f). However, BASF later became stricter in an attempt to guard both the progress made during the later stages of technological development and pending patent applications (Stoltzenberg 1994, pp. 165–168), (Szöllösi-Janze 1998b, pp. 181–183). Confidentiality was of even greater concern after Mittasch demonstrated in January 1910 that his composite catalyst produced results comparable with osmium and overcame a critical hurdle toward industrialization (Mittasch 1951, pp. 91–121), (BASF 1910a,d). As a result, the last report by Haber and Le Rossignol on the subject of ammonia synthesis from the elements was prepared in 1909 but did not appear until 1913 (Haber 1909c; Haber and Rossignol 1913b).

Haber wanted to publish his results to defend himself against Nernst's comments, which had continued in literature into 1910 (Haber 1909e, 1910d; Nernst 1910). But he also wanted to publicize his own success as a scientist. In January, he requested permission from BASF to publish his results on ammonia synthesis as a lecture, just as he had done for the results on the electric arc (Haber 1910a,d,h). It was important to Haber to introduce the public to the ammonia synthesis process as a whole. Other factors, even details of the catalysts, were of secondary importance because negotiations with BASF over the release of such details would be a lengthy process. Haber also argued it would benefit the patent process (Haber 1910f). Although Bernthsen was wary of losing the edge BASF had on the competition as well as possible adverse effects on the patent applications, he finally agreed to the publication of a shortened overview (BASF 1910h,k).

The negotiations on a final agreement, however, lasted through February and did not proceed without pressure from Haber. One of his initial drafts was too detailed

---

[5] "…im wesentlichen polemischer Natur…und dazu dienen solle um gegenüber den Angriffen des Herrn Professor Nernst recht zu behalten."

for BASF; they preferred he focus on the general fact that "the technical synthesis of ammonia from the elements, which heretofore had been assumed hopeless. . .has succeeded. . ."[6] Haber acquiesced at first, asking only for permission to send his detailed account to the Academy of Sciences in Vienna for later publication and, presumably, to secure his claim to the results (BASF 1910k; Haber 1910b). But in January of the following year, he notified the *Zeitschrift für Elektrochemie* of his lecture and planned a date with Carl Engler in Karlsruhe. Three weeks before the scheduled date, Haber had still not received permission from BASF. The two parties finally agreed on the content of Haber's lecture only days before he was due to speak (BASF 1910b,c,e,j; Haber 1910c,e,g,j). Haber held the lecture on March 18, 1910 in front of the *Scientific Society of Karlsruhe*.[7] BASF was still uncomfortable printing all the information Haber had presented at the lecture and insisted on an abridged published version. Haber framed his discovery within the context of agriculture and the explosives industry before focussing on the use of unusually high pressures and an osmium catalyst. "Through this result," he wrote, "the basis for an industry of synthetic ammonia is secured."[8] The iron and alumina catalyst received no mention (BASF 1910f; Haber 1910l,m), but the lecture and the publication produced the publicity Haber wanted. The event was covered in local newspapers, trade journals and commentaries, and Haber was contacted by members of academia, industry, and other personalities. The attention caught BASF off guard (BASF 1910g).[9]

In the context of Haber's success at BASF, it is also informative to consider Nernst's publication on the specific heat and equilibrium of ammonia from February of 1910 (Nernst 1910). In it, Nernst introduced new evaluations of the thermal characteristics of ammonia and vehemently defended Jost's 1908 results and their agreement with his 1906 heat theorem. The final page of the six-page publication was primarily devoted to the "polemic" with Haber, which Nernst blamed mainly on his peer. Nernst's tone in the description of the interaction, indeed his rehashing of the entire chronology on a scientifically factual, but also dramatic basis, shows the effect the episode had on him. It is also clear Nernst still viewed ammonia equilibrium as an important intersection between experiment and theory: it would aid in the verification of his heat theorem, which ultimately became a lengthy process (Nernst 1921, p. 359). Haber, on the other hand, discussed Nernst in private communication but not in his latest publication, the printed version of his 1910 lecture (Haber 1909e, 1910e,f). To some extent, Haber had moved on. Developments at BASF continued with Carl Bosch and Alwin Mittasch's successful conversion of Haber's apparatus into an industrial process; Haber himself already viewed the

---

[6] ". . .dass die bisherige als aussichtslos angenommene technische Darstellung des Ammoniaks aus den Elementen. . .gelungen sei. . ."

[7] *Naturwissenschaftlicher Verein zu Karlsruhe*

[8] "Durch diese Arbeitsresultate scheint die Grundlage für eine Industrie des synthetischen Ammoniaks gesichert."

[9] For further details see the correspondence at the Archiv der Max-Planck-Gesellschaft, Abt. Va, Rep. 0005: HS 2087, 2089, 2090, 2091, 2093, 2095, 2097.

Fig. 12.3 Berlin at the time Haber was being considered as director of the KWI for Physical Chemistry and Electrochemistry. By David Vandermeulen from *Fritz Haber: Les Héros* (Vandermeulen 2007); © David Vandermeulen/Guy Delcourt Productions

fixed nitrogen industry as secure, and his work in the field was coming to an end. Before the First World War, he oversaw further laboratory measurements for the ammonia system, experiments that increasingly tended toward basic research, but was less involved in Bosch and Mittasch's continuing work at BASF because of his confidence in the company and the abilities of its employees (Haber 1914a). In 1911, Haber became director of the KWI for Physical Chemistry and Electrochemistry in Berlin where he turned to other fields of research, including radiation, electron dynamics, and his old passion for the electric arc (Fig. 12.3). He would not return to the subject of ammonia synthesis (Haber 1911), (Coates 1937, pp. 1664–1670), (Mittasch 1951, p. 80), (Burchardt 1975, pp. 98–100), (Johnson 1990, chapter 7), (Stoltzenberg 1994, p. 170), (Szöllösi-Janze 1998b, p. 180), (Steinhauser et al. 2011, pp. 10–25).

Haber and Nernst were not diametrically opposed to one another's viewpoints, despite sometimes having different notions of what it meant to arrive at the correct solution. Nernst focused on scientific foundations, whereas Haber prioritized application and industry. Haber believed the end result was the description of a process that could be industrially upscaled; Nernst thought it to be agreement between experiment and theory (though he was also drawn to industrial implementation). However, they must have seen eye to eye on some things: Nernst supported Haber and BASF in their 1910 patent disputes on ammonia synthesis against industry competitors (Suhling 1972), (Stoltzenberg 1994, pp. 170–180), (Szöllösi-Janze 1998b, pp. 181–185), and Nernst helped convince Haber to accept the position of director of the KWI (Haber 1910k), (Stoltzenberg 1994, p. 206), (Szöllösi-Janze 1998b, p. 222).

"I have always strived toward the truth,"[10] Nernst is quoted as having said near the end of his life (Haberditzl 1960), (Bartel 1989, p. 109). These words, taken with a grain of salt, describe Nernst's approach to ammonia, just as it does Haber's.

Perhaps this unique intersection of ideas is what made them such productive rivals.

As for the emergence of Haber and Le Rossignol's detailed results on ammonia synthesis from the elements, their final report, "Über die technische Darstellung von Ammoniak aus den Elementen" (On the technical production of ammonia from the elements), was published in 1913 along with a report of a lecture Haber held at the Meeting of the German Chemical Society after BASF's patents were granted and construction of the plant at Oppau was in an advanced stage (Haber 1913a; Haber and Rossignol 1913b).[11] The next year, another short article appeared along with an extensive set of reports documenting refinements as well as new measurements carried out in Karlsruhe and Berlin, and, yes, in which the debate with Nernst continued (Haber 1914b,c; Haber and Greenwood 1915a; Haber and Maschke 1915b; Haber and Tamaru 1915c,d; Haber et al. 1915e,f). By then, the Haber-Bosch plants at Oppau and Leuna were in operation and poised to upend the world's nitrogen market, revolutionizing how humans produce nourishment and lead their lives. There was, however, still much to be understood from the scientific perspective.

---

[10] "Ich habe stets nach der Wahrheit gestrebt."

[11] August Bernthsen also presented Haber's process in New York City in 1912, but only after he was convinced the information he presented would have no effect on pending patents (Bernthsen 1913), (Bernthsen 1925, p. 52).

Chapter 13
The Role of Physical Chemistry as a Theory

Physical chemistry was a powerful, central tool in the discovery of ammonia synthesis. However, this does not mean the full potential of the theory was (or needed to be) employed to solve the problem. By studying the use of the theory in Haber's and Nernst's work, we are able to examine the role physical chemistry played in terms of theoretical and mathematical possibilities versus practical, industrial application. These considerations are based on the preceding text as well as the derivations in Appendices A and B, but are presented here with minimal mathematics. As before, a consideration of the appendices will improve understanding of the following discussion.

The key to the success of physical chemistry is contained in the use of the free energy, A, in Eq. B.1, as a quantitative measure of describing chemical reactions. Here, more background is provided to illustrate its role in Haber's and Nernst's work as well as the extent to which a precise expression of its value was needed. This viewpoint also frames ammonia synthesis in a new light: as part of the completion of classical thermodynamics (and, therefore, "classical" physical chemistry). What would later become known as the third law of thermodynamics is considered to have been the final step in this development (although it was not apparent at the time Nernst published it) (Suhling 1972). However, a theory itself is not complete without experimental substantiation. This complementary dynamic is what took place between Haber and Nernst throughout their exchange as ammonia research supplied meaningful data for the confirmation of Nernst's heat theorem. The discrepancies Nernst observed between theory and experiment could have been based on behavior in nature that had not yet been discovered or due to an error in his assumptions or calculations. Haber and Nernst's interaction showed, at least for the ammonia system, that the deviations were caused by uncertainties in thermal data and not by the physical/mathematical descriptions of the free energy.

The origin of these descriptions lies in research on the development of heat during chemical reactions. Depending on experimental conditions, the free energy is referred to as the Gibbs energy or the Helmholtz free energy, named for Josiah Willard Gibbs and Hermann von Helmholtz, who independently derived expressions

© The Author(s) 2022
B. Johnson, *Making Ammonia*, https://doi.org/10.1007/978-3-030-85532-1_13

and published them in 1878 (Gibbs 1878a,b) and 1882 (von Helmholtz 1882).[1] The "isolation" of the free energy as a fundamental value arose from attempts to assess chemical affinity through the heat released during a chemical reaction. In the first half of the 1800s, there were many investigations into this phenomenon (Nernst 1914), (Partington 1964, pp. 608–620, 684–699), (Suhling 1972), (Bartel 1989, pp. 71–73). By mid-century, quantitative attempts to determine the maximum work available from a chemical reaction of homogeneous substances (for example the electromotive force (E.M.F.) that could be expected from an electrochemical cell) had resulted only in the unsatisfactory assumption that the heat of reaction was equal to the maximum work (and a measure for the chemical affinity). That is, $\Delta U = \Delta A$, with the spontaneous chemical reaction generating the most heat. These ideas were (at least partially) supported by James Joule, Julius Thomsen, Marcelin Berthelot, Helmholtz and others and formulated in different ways under various names, including the *Principle of Maximum Work*. Despite its simplicity, the general validity of the claim was viewed with skepticism as specific chemical systems behaved in obvious contradiction. Some chemical reactions provided an E.M.F. while proceeding endothermically (ΔU was negative) and there were also indications the heat of reaction was temperature dependent. With the convergence of chemistry and energy science, it became evident that there were two kinds of energy involved in a chemical reaction. The quantities were, in general, not equal and their magnitude depended on reaction conditions. One was always in the form of heat and the other could be obtained partially or fully as electrical energy or as another form of work. A chemical reaction proceeded spontaneously in a way that minimized the latter, the free energy.

It is a temperature-dependent dynamic that describes a balance in nature:

$$Free\ Energy = Total\ Energy - Temperature \times Entropy \tag{13.1}$$

where *Total Energy* may denote either the internal energy or the enthalpy. At high temperatures the (*Temperature* × *Entropy*) term dominates, and the free energy is minimized through reactions maximizing entropy. At low temperatures, this term becomes small and reactions occur that minimize the *Total Energy*. At non-extreme temperatures, nature strikes a balance between the total energy of the system and the product of temperature and entropy to minimize the free energy at equilibrium (Müller 2007, pp. 148–149). But ammonia has one more trick up its sleeve, which was the "mystery" that plagued researchers during the nineteenth century (see Part I, Chap. 4). In a mixture of NH_3, N_2, and H_2 at 300 K and 1 bar, the equilibrium conditions favor a nearly full conversion to ammonia as the decrease in total energy during the NH_3 production outweighs the decrease in entropy (the latter on its own would hinder the exothermic ammonia generation). However, no reaction occurs due to the potential barrier from nitrogen bonds (as well as from hydrogenation and

---

[1] These are commonly discussed derivations of the free energy, but they were not the first (Partington 1964, pp. 614–616).

blocked active catalyst sites) (Vojvodic et al. 2014). If the temperature is raised in order to overcome the energy barrier, the *Temperature × Entropy* term also becomes large. The change in this value must then be countered by increasing the pressure, which in turn lowers the entropy. The catalyst facilitates the reaction because it reduces the temperature required to overcome the energy barrier. It also enables the use of lower pressure to still achieve an adequate ammonia yield (Müller and Weiss 2005, pp. 63–71), (Müller 2007, pp. 156–159), (Müller and Müller 2009, pp. 269–272). Thus, ammonia synthesis is described as a "balancing act." This was the crucial knowledge that had been established by physical chemists, and was well-known to Fritz Haber and Walther Nernst as they began their work on ammonia synthesis (Haber and van Oordt 1905b, p. 343).

While Helmholtz' definition of the free energy supplied decisive information about the behavior of chemical systems, it only allowed for a partial determination of the defined quantity. The differential equation still needed to be integrated, a mathematical operation that necessitates the addition of an unknown constant to the final equation (Appendix B, Eqs. B.1 and B.2). The *change* in chemical equilibrium with temperature could be calculated but the *absolute* value could not. By the time Haber and Nernst took up the problem of ammonia synthesis, the determination of this constant (and, thus, of chemical equilibrium) had become a central challenge in physical chemistry. "The problem," wrote Lothar Suhling, "which faced the few chemists in the nineteenth century who were dealing with the application of thermodynamical principles on chemical equilibrium, was the old question of quantitative consideration of chemical affinities (Suhling 1972)." At the time of Haber's breakthrough, there were two methods of solving this problem. Usually the chemical equilibrium (or alternatively the E.M.F) was measured directly, and the data point was used to determine the constant of integration (Nernst 1914, 1921). This method solved the problem in principle, but left the resulting values of chemical equilibrium subject to experimental error on the order of magnitude of any subsequent measurements. It defeated much of the purpose of the application of theory.

Walther Nernst's goal was to find a way around this empirical determination of the constant, and with the help of a new assumption about nature, he succeeded. His theory was published in 1906 (Nernst 1906), and to be sure, the chemical industry recognized the value of such a theoretical tool: chemical yields could be determined without having to carry out the (possibly costly) reaction itself (Haberditzl 1960; Suhling 1972).

Nernst's solution stated that the equilibrium of chemical systems depended on a specific behavior at temperatures approaching absolute zero. He had questioned the universal applicability of the principle of maximum work ($\Delta U = \Delta A$) since his investigations of chemical equilibrium and heats of reaction in the 1890s in Göttingen. The concept could not be universally applied and Nernst was convinced the resolution of the problem would be a profound discovery (Haberditzl 1960; Nernst 1914, 1921), (Bartel 1989, pp. 21–39). "A hidden law of nature," he wrote in the first edition of his *Theoretische Chemie* (*Theoretical Chemistry*) in 1893, "underlies the 'principle of maximum work' whose clarification is of highest importance (Nernst 1893, p. 543)."[2] He was not alone in his suspicion. Others such as Le Chatelier and Fritz Haber himself (Haber 1905b, pp. 41, 50, 62–64) were considering the physical meaning of the integration constant. Nernst openly admitted that Le Chatelier had already formulated the problem clearly in 1888 (Chatelier 1888), but it was left to Nernst to present the answer. It would later became known as the third law of thermodynamics, though at first, it was only a postulate based on a hunch he had had after considering experimental data (Nernst 1906, pp. 5–7), (Haberditzl 1960, p. 408), (Farber 1966, pp. 185–186), (Hermann 1979, pp. 131–132), (Suhling 1993), (Müller 2007, pp. 170–172).

Therefore, after 1906, both a theoretical and a (theory-based) experimental solution were available to determine chemical equilibria. How does this situation help us identify the aspects of physicochemical theory that were necessary for Haber's breakthrough?

If we consider Haber's and Nernst's results in Figs. 11.7, 11.8, 11.9, 11.10, and 11.11 and compare them to the supplied modern theoretical values, we see that Haber's 1905 results were highly accurate considering there was no information available about the equilibrium between ammonia, nitrogen, and hydrogen at the time. By 1907, Haber's accuracy would not (and did not need to) be further improved. He achieved his results not by using Nernst's theory, but rather his own approximations for chemical equilibrium (Appendix A) combined with an experimental measurement. Complete knowledge of the theory of physical chemistry was, therefore, not needed to achieve the breakthrough. To be exact, Haber's approximations were only accurate within a narrow temperature range surrounding the experimental data point (the deviation can be seen in the graphs). However, this local accuracy puts the current discussion into stark relief. The approximations were not perfect enough for a full description of the system, but they were powerful enough to achieve the intended objective: industrial upscaling. If, however, the goal had been to achieve full knowledge and control of the chemical system under arbitrary conditions, it would have been another matter.

Stepping back to consider the state of knowledge of the free energy before theories such as Helmholtz' appeared, there was not enough information to allow for a *controlled* synthesis of ammonia under any conditions. I use the word "controlled" because it would, of course, have been possible (although unlikely) for someone to stumble upon the correct conditions. As illustrated above, equations like Helmholtz' fostered the idea of the balancing act (Eq. 13.1) and, thus, a precise way in which conditions must to be changed in order to achieve new equilibria. Without this knowledge, even if someone had serendipitously achieved a synthetic production of ammonia under some set of conditions, it would not have been possible to extrapolate the suitable conditions needed for industrial synthesis (especially considering that these conditions were not technologically attainable in the nineteenth century).

---

[2] "Dem 'Prinzip der maximalen Arbeit' liegt ein Naturgesetz versteckt zu Grunde, dessen weitere Klarstellung höchste Wichtigkeit besitzt."

Thermodynamic derivations of the free energy, such as those provided by Helmholtz or Gibbs, combined with Haber's approximations and measurements (including his results on the effect and correct role of the catalyst) were the critical pieces of the physicochemical puzzle needed to achieve the breakthrough of ammonia synthesis. The importance of these combinations of knowledge is underscored by the failure of the late empirical approaches, like Edgar Perman's. Even basic physicochemical principles, such as Le Chatelier's principle, were not critical to Haber's initial experiments, although they certainly influenced his thinking and were critical to the subsequent upscaling. Nernst's theory provided a common starting point for the exchange of knowledge between Haber and Nernst as well as a solid theoretical framework for chemical synthesis and reassurance for industry. While such a framework that included all physicochemical principles was not required for industrialization, it would inevitably garner great interest as it was indispensable to securing scientific knowledge at a fundamental level.

It is important to note that this circumstance, while determined by the particular skills of those involved, was also shaped by the state of research at the time. A stronger contribution to the breakthrough from theory was not an impossibility. In 1906, Max Planck formulated an expression for entropy based on the probability interpretation of Ludwig Boltzmann and others that specified a zero-point for S

Fig. 13.1 From left: Walther Nernst, Albert Einstein, Max Planck, Robert Millikan, and Max von Laue in 1931, probably at the home of the latter. *Source*: Archive of the Max Planck Society, Berlin-Dahlem, III. Abt. Rep. 57, NL Bosch; Akz 46/95, Picture Number III/2

at $T = 0$. From this reference, it was possible to derive Nernst's heat theorem and provided the third law of thermodynamics with some theoretical underpinning. However, Planck's approach did not convince everyone, among them Albert Einstein, and the correct physical starting point to arrive at his expression of entropy was debated for some time (Planck 1906, pp. 129–137), (Grandy 1987, pp. 16, 74), (Kox 2006) (Fig. 13.1). Had such a rigorous mathematical formulation allowed for an earlier substantiation of Nernst's theorem, his theoretical determination of ammonia equilibrium—one that would have relied on the full theory of physical chemistry— may have been considered complementary to Haber's initial measurements. At the time though, Nernst's theory contained too much conjecture. Later, when the matter was settled and his theorem had proven correct and applicable the impact was clear: the field of classical thermodynamics was complete.

Chapter 14
Further Reflections on Scientific Discovery and The Haze

We have followed the scientific developments of ammonia synthesis from the elements from the initial determinations of equilibrium through to the beginning of industrial upscaling during the first decade of the twentieth century. A central result of this presentation is that natural scientific details about the involved chemistry and physics could be accurately presented. Not only does the combination of scientific and historical context to Haber's work clear up misunderstandings about the scientific developments and Haber's interaction with Nernst, it also allows us to understand why Fritz Haber took on the challenge of ammonia synthesis in the first place (Chap. 10). His attraction stemmed largely from his interest in fundamental principles, although he was not as preoccupied with this aspect as Nernst. Haber's enthusiasm should not, however, be completely decoupled from possible industrialization of new technology, financial gain, or the need for the production of raw materials for fertilizer and explosives. The set of circumstances as well as the scientific arena in which he acted was complex and improbable. It reminds us of the reality of scientific discovery: the time must be right, and even then, biases or preconceived notions can impede progress (Haber 1920, p. 326), (Kuhn 1959). But the obstacles are only temporary.

The central historical contribution (and the most tangible example for the progress of science) was the "interaction" or "exchange" between Haber and Nernst, a dynamic which extended beyond the meeting of the Bunsen Society in Hamburg in 1907. Nernst, whether consciously or not, incited Haber to further his work on ammonia on three distinct occasions apart from the involved and sustained public debate in literature: the letter in 1906, the Bunsen Society meeting in 1907, and the meeting in 1908 in Berlin where Haber learned of Nernst's relationship with Griesheim-Elektron. The development displays many random events with respect to future outcomes. In retrospect, however, it is clear how a basis in fundamental physical principles and mathematics leads to success. Context for their exchange was also provided by Haber's and Nernst's individual approaches. Haber and his team learned the importance of the verification of theory through experiment while Nernst and his co-workers understood the power of experimental results to

© The Author(s) 2022
B. Johnson, *Making Ammonia*, https://doi.org/10.1007/978-3-030-85532-1_14

determine something which could not be described by theory: the behavior and efficacy of a real-life catalyst. Haber and Nernst's interaction provided a fact-based, alternative objective (scientific progress) when it appeared physicochemical realities may not allow an industrial upscaling. Money and recognition also played their part. While the meeting of the Bunsen Society may have been the apex of the exchange with its element of drama, their ongoing relationship as a whole must be considered the important factor in the successful conclusion to ammonia synthesis from the elements.

It is important to reiterate that at some point after the Bunsen Society meeting, Haber's sense of certainty toward his own results crystallized, despite Nernst's continued commentary. Lawrence Holmes wrote (in reference to Lavoisier but still fitting here) that a development like Haber's toward increased confidence is a consequence of the "conditioning effect of the passage of time [...] At some point. . .a scientist must make the decision to commit himself personally to what he has come to believe, even though he has not resolved all of his own doubts, and can never be certain he is right (Holmes 1985, p. 107)."

Further considering the developments after the events in Hamburg, Friedrich Jost's own theoretical and experimental abilities became apparent through his independent publications. It was not only Haber and Nernst who were able to produce these studies. In consideration of the assistants and technicians van Oordt, Le Rossignol, Kirchenbauer, Jost and Jellinek, the base of individuals who contributed to the scientific breakthrough broadens and moves us away from the single story of one courageous scientist devoting himself over years to the advancement of a narrow sliver of knowledge. In fact, Haber's story alone challenges this stereotype: he actively helped his assistants Heinrich Danneel, Friedrich Kirchenbauer, Gerhardt Just, Adolf König, and Robert Le Rossignol navigate the academic world and was consistent in recognizing and publicizing the contributions of others, including the Margulies brothers (BASF 1910i; Haber 1909d, 1910i; Krassa 1955). He cared deeply about research and education and was known for his fervent scientific discussions with colleagues and students (Engler et al. 1909; König 1954; Krassa 1955; Schlenk 1934), (Sheppard 2020, p. 53) (Fig. 14.1). Even politically, Haber showed deftness in his position at the university, in his negotiations with BASF, in the establishment of the Kaiser Wilhelm Institute for Physical Chemistry and Electrochemistry, and in his relationship with Nernst. He was not a one dimensional man fixated on a singular scientific goal, but exhibited growing variety in his academic roles as professor, researcher, and mentor.[1] Nernst (and Ostwald) exhibited similar traits with their students and in their careers (Bartel 1989, pp. 34, 51). Perhaps a more accurate, albeit unwieldy name to consider is the *Haber-Bosch-Mittasch-Nernst-Ostwald-Le Rossignol-van Oordt -Kirchenbauer-Jost-Jellinek-Process.*

---

[1] I have intentionally left out any mention of Haber's private life. The consequences of his behavior are for the interested reader to assess themselves in any number of sources.

Fig. 14.1 "Parody Photo: The Colloquium." At the Institute for Physical Chemistry and Electro-chemistry, Technical University of Karlsruhe in 1909. There are many indications that Haber's staff was not "all work and no play." *Source*: Archive of the Max Planck Society, Berlin-Dahlem, Jaenicke and Krassa Collections, Picture Number VII/5

While my study of the history of ammonia synthesis has brought me to these particular conclusions, it has also left a central question unanswered: why was Haber at times aloof, or even dispassionate toward the nitrogen-hydrogen-ammonia system during the years he was intensely occupied with it? It began with the matter-of-fact way he and van Oordt presented their strikingly accurate 1905 results in a field where so many had tried and failed before them. What they had achieved was momentous and symbolic, but Haber initially made little of it, perhaps because industrialization appeared unlikely. Furthermore, it is puzzling why Haber never discussed the scientific side of ammonia synthesis in his correspondence. He discussed an array of other topics in detail and with obvious fervor. Did such letters simply not survive? Or did the certainty he had in the accuracy of his measurements render such discussion superfluous? This interpretation fits well with the increasingly confident tone found in his publications and, as I put forward in Chap. 10 in consideration of his wider scientific interests, that he saw ammonia mainly as a vehicle to apply and complement the theory he had already developed. The first set of measurements from 1905 accomplished this goal and perhaps there was nothing more he wished to achieve. Despite Haber's tendency in Karlsruhe to move toward more fundamental problems, in the case of ammonia synthesis he never seemed to reach Nernst's level of contemplation with regards to the basic physical mechanisms behind the chemical dynamics. This circumstance would also help explain why he never returned to the subject after the upscaling at BASF. In a nutshell, it is not clear what exactly Haber's intention was after he published his initial 1905 results with van Oordt and his scientific goals had apparently been reached. It was more than money or reputation and may have changed over time.

Perhaps the fact that Chap. 10 contains such a wide array of factors explains the ambiguity.

Whatever the reason, it is curious that the (arguably) most important piece of research of Haber's career occupied him in a such narrow manner, both in terms of time and intellectual curiosity.

Returning to the concept of *The Haze*, the objective of Part II was also to show, beyond a discussion of the scientific details, that the arena in which discovery takes place is complex and includes any number of elements. While this characteristic is often postulated, here we can discuss concrete causes. In the introduction to Part I, I alluded to the theoretically unlimited set of influential factors that renders this concept of little practical value. Rather, it is useful to consider how this complexity has been illuminated. What are the key components that forged it and how are they interlocked? The historical run-up to 1903 described in Part I also possesses a convoluted dynamic that will result in a mature scientific arena at an uncertain point in time, under uncertain circumstances. Again, the time must be right. However, the events of Part I may be the more simple set of complexities. Once the stage has been set for a Fritz Haber, a Walther Nernst, and their assistants, the "microcomplexities" within the mature arena begin to emerge (Kanter 1988; Sgourev 2015). In the case of ammonia, the complexities were increased by Haber's and Nernst's unique backgrounds suspended between experiment, theory, and industry, the technical prowess of the assistants within the context of the then-technically achievable, outside financial incentive provided by industry, political, and societal pressures, the state of theoretical knowledge, and efficacy of applicable approximations. Different forms of knowledge exchange along with political wrangling contributed to the evolution within this setting. Considering these intricacies, the reasons for the complexity is no longer a surprise. Increased attention to local, national, or global circumstances, movements of knowledge, or the effects of the academic setting could extend the detail even further (Renn 2012; Wendt 2016).

We may also consider the resolution of the problem. The mystery of ammonia synthesis has been solved; we have been able to synthetically produce the chemical for over a century. But are we, after having reviewed the subordinate "microcomplexities" in this section, able to clearly identify the scientific breakthrough? Was it Haber's initial 1905 measurements? They were precise enough to form the basis of further scientific and industrial advancements. Was it one of the subsequent publications with increased accuracy or knowledge of improved technical possibilities (i.e. higher working pressures)? Or the Bunsen Society meeting in Hamburg? Or was it later at Haber and Le Rossignol's 1909 demonstration in which all of the elements needed for industrial synthesis were first brought together?[2] Perhaps it was in 1910,

---

[2] Smil states that this is the moment of the "decisive breakthrough" (Smil 1999), (Smil 2001, p. 81). Sheppard, highlighting the ambiguity of a scientific discovery, qualifies Smil's statement with reference to Haber's preceding years of laboratory work (Sheppard 2020, p. 157). Rudwick, commenting on the moment of resolution in the Devonian Controversy, writes that "as soon as conflict and controversy within the core set [of actors] for any focal problem are replaced by virtual consensus, the focal problem is at an end...(Rudwick 1985, p. 427)." He continues, however,

when Haber gave his lecture in front of the Scientific Society of Karlsruhe and made the breakthrough public for the first time. Or was it a combination of all these, extended in time?

The answer may hinge on our perspective or how we frame the question and depends on our definition of the terms "breakthrough" or "discovery." Are the meanings of these words concrete enough that we truly understand each other when we use them? If not, we may need a finer nomenclature to better characterize the progression of science. Such an improvement would certainly aid in understanding the dynamic of the Haze.

---

that the "controversy came to an end…not when some set of new theoretical concepts had been formulated, nor when some set of new observations had been made…As a matter of history, the controversy ended as soon as a set of less than ten leading geologists…converged in a collective judgement that the problem *had* received a satisfactory solution and that a reliable new piece of natural knowledge had been shaped [p. 428]." In the case of ammonia synthesis, the sense that an end had been found was not simultaneously shared by everyone. After Haber and Le Rossignol had determined their "satisfactory solution" in 1908, Nernst and Jost remained combative.

Part III
The Haze:
A Theory for Breakthroughs in Science

Up until this point, we have examined in detail the breakthrough of ammonia synthesis from the elements. The story has many facets and has been used (and will be used again) as the basis for historical, scientific, technological, moral, gender, environmental, and military discussions. The primary goal of this book is to provide a science centered narrative of ammonia synthesis missing from existing historical literature, but it also lays the groundwork for a general description of science based on the interaction between scientists. To achieve this second goal, I draw on the case study presented in Parts I and II and also on my time actively working as a physicist researching solid state systems (semiconductors for photovoltaics and noble metal oxide catalysts), both theoretically and in the laboratory. While my analysis makes use of a wide range of concepts and terminology from different disciplines, including social network and innovation theory, as well as science philosophy, I am not trained in these fields. I approach the challenge of generalizing our view of a scientific discovery mainly as a physicist, but also as a historian of science. I make no deliberate attempt to go beyond this standpoint, just as throughout the book, I have limited the narrative to science. This narrowed perspective does not make the resulting lessons incomplete. Rather, the approach provides a building block for non-scientists to draw upon if they wish to expand the scope of their own views on scientific research and discovery. My hope is that the following description, though not exhaustive, will provide all who are interested with a stimulating and provocative look at the incomparable, yet wholly human scientific endeavor.

Chapter 15
Terminology

In Parts I and II, the scientific breakthrough of ammonia synthesis was described in terms of a confluence of factors leading to an arena of research mature enough for advancement—it is a dynamic encapsulated in the concept of *The Haze* (Schlögl 2018).[1] In this particular arena, scientists determined the conditions under which ammonia can be synthesized. The upscaling of the Haber-Bosch process at BASF followed, along with its proliferation and secondary effects, such as the establishment of the high pressure catalytic industry.

Here in Part III, we go into theoretical detail on the structure and dynamic of the Haze. It is a "manual" approach, which can be viewed as complementary to computer-based historical methods and studies that analyze the dynamics of larger or global groups (Fangerau 2010; Graßhoff 1994, 1998, 2003; Graßhoff and May 1995; Renn 2012; Wendt 2016; Wintergrün 2019). In particular, I wish to discuss the nature of the individual interactions leading to knowledge exchange in science and how this activity enables scientific progress by simultaneously exploiting several paradigms. The significance of these events, often without a plan or strategy, may only be recognized by the small number of people involved, if at all. While these interactions do not influence the outcome of scientific research, they are fundamental to the fact that science works at all and their random, stochastic nature succinctly explains the complexity of the Haze. The different approaches to investigating the outcome of many such interactions focus on either large or small groups of actors and share a reliance on case studies. While the reactions to this strategy are varied (Schumpeter 1942, p. 83), (Rudwick 1985, pp. xxii, 15–16), (Basalla 1988, p. 30), (Siggelkow 2007), in many examples, case studies have been shown to have great illustrative power (Fangerau 2010; Fleck 1980; Globe et al. 1973; Holmes 1985; Padgett and Ansell 1993; Padgett and McLean 2006; Padgett and Powell 2012b; Rudwick 1985; Sanderson and Uzumeri 1995; Sgourev 2013, 2015), (Obstfeld 2017, Chapters 4, 5). The case study of ammonia synthesis, too,

---

[1] The concept of the Haze was developed with significant input from Prof. Dr. Robert Schlögl at the Fritz Haber Institute of the Max Planck Society. He also suggested the term "Haze".

© The Author(s) 2022
B. Johnson, *Making Ammonia*, https://doi.org/10.1007/978-3-030-85532-1_15

offers more than only a recounting of events; generalities of science can also be considered. The Haze is a complex phenomenon and the relatively clear-cut nature of the breakthrough of ammonia synthesis suggests a simplified schematic with which we can discuss the revealing of new knowledge in science and its movement into technical and commercial domains.

For practical reasons, continuity, and an attempt at a common lexicon, I have used terminology from social network theory and innovation literature to generalize the dynamics of change in science. Researchers in these fields have had success illustrating the mechanisms of innovation in diverse disciplines but application to purely scientific episodes is rare. In employing their tools, I wish not only to stretch the concepts to cover scientific breakthroughs but also to enable the description of the transfer of knowledge resulting from basic research. There is no attempt, however, to develop these theories further. I also make use of terminology from philosophical perspectives on science and technology from throughout the twentieth century while trying to remain concrete about the mechanisms of progress in the mature, exact natural sciences as they exist today.

The transfer of terminology is not without consequence: the approach uncovers facets of scientific activity that differ from other processes of innovation and in doing so, provides vivid examples of how science progresses toward a discovery. This difference is not the result of the primacy of science or scientific progress over, say, technological progress; technology provides the novel instrumentation and techniques to meet the needs of ever more sophisticated scientific research (Brooks 1994). Science and technology are intimately linked and dependent on one another (and are better off for it) (Basalla 1988, pp. 27–28), (Rogers 1995, pp. 140–141), (Bonvillian 2014). It is rather that the discussion here centers on science because literature on scientific progress remains sparse and because science is my area of expertise. This description is not of the "scientific method," which is essentially based on honesty of thought, the forthright presentation of results, and the acceptance only of testable ideas which may be disproved. Rather, it is an examination of scientific research as an activity within the framework of existing theories of change that gives tangible descriptions of its driving mechanisms.

Within the terminology of social network and innovation literature, there is widespread use of physical, chemical, biological, etc. (in short: natural scientific) concepts and terms to imply analogy between processes. My initial reaction was to reject such usage. One example that is critical within the context of physical chemistry is the term "catalyst" used in the sense that networks and/or novel combinations of knowledge can have an outsized impact. While this description embodies some of the properties of a catalyst, it should be compared to the definition given by Ostwald in 1902: "A catalyst is any substance that, without having an effect on the end product of a chemical reaction, changes its speed [reaction rates] (Ostwald 1902)."[2] While a catalyst often has an oversized effect compared to the

---

[2] "Ein Katalysator ist jeder Stoff, der, ohne im Endprodukt einer chemischen Reaktion zu erscheinen, ihre Geschwindigkeit verändert."

quantity of material present, it is not required to behave this way; it is only required to change the reaction rates, often suppressing specific reactions while favoring others. The general mathematical term for this kind of behavior is "non-linear" (Quinn 1985).[3] Such mathematically-based scientific terms are strictly defined for exact application to experimental and theoretical work, and there is no tolerance for creative reinterpretation. As much as possible, there is no subjectivity left in the term. If something is missing, if a new term is needed, it may be derived and characterized. The task of defining terminology in the most unambiguous way possible has long been considered vital to the field, although the meanings of scientific terms can develop over time as their exactitude (hopefully) grows (Popper 1935, p. 11), (Rudwick 1985, pp. 401, 446–448). As I ventured into the theories of change and innovation, however, the power of the analogies became apparent along with the recognition that they had been used with much care (Basalla 1988; Brooks 1994; Langrish 2017; Laubichler and Renn 2015; Murmann and Frenken 2006; Obstfeld 2017; Padgett 2012a; Padgett and McLean 2006; Sgourev 2015; Ziman 2000). I became so convinced of their practicality, in fact, that I have used them myself in describing the Haze. For maximum benefit, the use of descriptive terms from other fields should be accompanied by a concerted effort to understand their origin and original meaning.

It will be no surprise that I have received pushback on my own use of certain terms. One example is the distinction between *discovery* and *innovation*. Another is the use of the word "science" as if it were well-understood to mean physics, chemistry, biology, geology or any of the natural or "hard" sciences. It is not clear to everyone, however. Yet continuously attaching a prefix is cumbersome so that after this declaration, I have decided to leave it off in the follow discussion.

---

[3] What is meant is "non-linear" with the exponent greater than 1.

Chapter 16
Normal Science

The starting point for the discussion of scientific discovery is Kuhn's distinction between "normal science," in which routine research is performed and the moments of crisis which can lead to paradigm shifts (Kuhn 1970, Chapters 3 and 4).[1] We begin with a description of scientific discovery within the context of normal science—that is, within the context of a given, accepted paradigm—before expanding to a consideration of an extra-paradigmatic dynamic. Working within a paradigm means, among other things, that there is an accepted "objective" set of methods, concepts, theories, and related expectations to which new experimental and theoretical results can be compared. While theoretical results tend to be more succinct, experimental results extend over a range of possibilities. To narrow the experimental field, I am specifically referring to the recorded reaction of a system either to an excitation (a signal) or to a specific set of conditions (a measured yield). In terms of ammonia synthesis within the framework of physical chemistry, if a specific amount of ammonia is sought, there is precisely one temperature at a given pressure that will provide this quantity from a given starting ratio of nitrogen and hydrogen. There is no strategy that will lead a scientist to another set of conditions that may be more beneficial to certain individuals for economic, political, or any other reason.

Among the activities of normal science, also called "fact-gathering" or "puzzle-solving," Kuhn described three main pursuits. The first is the determination with increased accuracy of "that class of facts that the paradigm has shown to be particularly revealing of the nature of things." These are measurable quantities, positions, characteristics, et cetera that are integral to the paradigm, for example, stellar position, electrical conductivity, chemical composition, or thermal properties. The second pursuit is the determination of "those facts that, though often without much intrinsic interest, can be compared directly with predictions from the paradigm theory." Put simply, this pursuit is the demonstration of agreement between theory

---

[1] Although Kuhn used the term paradigm in more than one sense, here I associate it with what was his "core element:" a scientific (often mathematically based) theory (Klein 2016a).

© The Author(s) 2022
B. Johnson, *Making Ammonia*, https://doi.org/10.1007/978-3-030-85532-1_16

and experiment. It was a decisive factor in the interaction between Fritz Haber and Walther Nernst. The third activity "consists of empirical work undertaken to articulate the paradigm theory, resolving some of its residual ambiguities and permitting the solution of problems to which it had previously only drawn attention." This activity, Kuhn argued, is the most important class of fact-gathering and can be broken down into three subcategories: the determination of physical constants, the determination of quantitative laws, and the application of the paradigm to other areas of interest. These activities are nearly the same for theoretical and experimental endeavors, especially paradigm articulation, though the amount of work devoted to each of the three classes of fact-gathering may differ.

To underscore that there is precisely one state of a physical system under defined circumstances, consider Kuhn's attitude toward a particular form of paradigm articulation.

> Few of these elaborate endeavors [to determine the value of universal physical constants] would have been conceived and none would have been carried out without a paradigm theory to define the problem and *to guarantee the existence of a stable solution* [italics added].

Returning to physical chemistry and the conditions for chemical synthesis, the accuracy to which we know the universal constants determines how well we can predict the outcome under our chosen state variables (see the constant R in the ideal gas law, for example). With these guidelines for identifying "normal" scientific activity, the case study of ammonia appears particularly appropriate. It is not only a clear-cut example of a scientific breakthrough, but it fulfills the criteria of normal science: the determination of physical constants (from thermal data), the comparison of experiment and theory, the resolution of ambiguities, and the formulation of quantitative laws.

Considering these elements of Kuhn's theory, we may pose an interesting question. The breakthrough of ammonia synthesis is just that: a breakthrough and not a paradigm shift in any Kuhnian sense. It is a rare moment that results in the identification of especially pertinent and applicable scientific knowledge, not only for subsequent scientific undertakings but also for technological and industrial progress with potential for societal and/or cultural impact. But there is no mechanism in Kuhn's description of "normal science," apart from the arrival of a period of crisis leading to a paradigm change, that leads to anything but eternal fact gathering. Yet breakthroughs do exist—we have proof. What is it, then, that differentiates a mundane laboratory assessment from a historical, world changing experimental result? Why will one publication, based on years of meticulous research, receive no attention, be cited zero times, and descend to the depths of the sea of scientific literature while another becomes the basis for an industrial process that can help feed a third of the human population currently alive, provide vast munitions for the war machine, all while making meaningful contributions to the completion of a physical theory?

As I have hoped to illustrate in Parts I and II, it is a question of knowledge, talent, and intuition, but perhaps most of all, of circumstances.

We now turn to the Haze to reformulate the relationship between normal science and paradigm shifts.

Chapter 17
The Structure of The Haze

A general structure of the Haze has already been suggested by the title of Part I: *A Confluence of Factors*. Both the temporal juxtaposition of events, which eventually, albeit at an unknown point in time, results in the arena for discovery, as well as the experimental and theoretical scientific details contribute in differing yet interdependent ways (Fleck 1980, p. 128), (Graßhoff 2008). There is no "grand design," neither in the minds of those ultimately responsible for the decisions nor in the outcome of individual experiments. It is an adaptive process during which we are led one step at a time. In reference to the rise of the Medici family in fourteenth century Florence, John Padgett, *et al.* wrote, "heterogeneity of localized actions, networks, and identities explains both why aggregation is predictable only in hindsight and how political power is born (Padgett and Ansell 1993)." If we replace "political power" with "a scientific discovery" the statement remains just as applicable.

There are many examples in the literature in which a dynamic is described with stages of progress, including basic scientific research and industrial and commercial development, that lead to technological innovation (Bush 1945a), (Rogers 1995, pp. 131–137), (Branscomb and Auerswald 2002). One classic description is Vannevar Bush's "pipeline," introduced in 1945 as a post World War II effort by the U.S. federal government to prepare basic research results for use in private industry. Another concept envisions a "valley of death" dramatizing the demands put on entrepreneurs during the shift from invention to innovation. A similar metaphor describes a chaotic "Darwinian Sea" inhabiting the divide between the shores of science/technological and investor/industry which only the most well-suited ideas may cross (Auerswald and Branscomb 2003; Bonvillian 2014; Branscomb and Auerswald 2002; Bush 1945b). A more intricate example is "the chain-linked mode of innovation," a reaction to theories describing the innovation process as having some kind of conceptual order in the hopes that clearer organization would aid in policy development. Here, multiple feedback loops exist between scientific research, technological innovation, product design and marketing as opposed to a linear process (Kline and Rosenberg 1986; Sanderson and Uzumeri 1995).

© The Author(s) 2022
B. Johnson, *Making Ammonia*, https://doi.org/10.1007/978-3-030-85532-1_17

The interdependence of scientific (or basic) research and technological innovation is rightly emphasized in many of these models, especially the more recent ones (and has been a point of criticism toward Bush's pipeline (Brooks 1994; Langrish 2017)). However, while some models of innovation view scientific research as the earliest possible step (one that is not always necessary or may be revisited through feedback loops), it is sometimes missing altogether, even in cases where it plays a role. Descriptions of the mechanism of scientific progress itself are even scarcer (Rogers 1995, p. 132). Rather, a successful scientific outcome is often used as a starting point, or at least presupposed. In itself, the point in time immediately following a scientific breakthrough is not an invalid place to begin a discussion of technological innovation, but a discussion of the preceding scientific progress would be beneficial because scientific discovery and technological change are inevitably intertwined. Unfortunately, attempts at an explanation can fall victim to the same oversimplification that innovation researchers themselves warn against when describing technological advances.

As was shown in the case of ammonia synthesis, normal scientific research is as continuous as innovation in technology and change in other fields. The notion is widespread but is not always applied to science (Schumpeter 1952, p. 89), (Globe et al. 1973), (Stigler 1982), (Becker 1984, pp. 301–310), (Chandrasekhar 1987, p. 14), (Basalla 1988, pp. 30–57). "The final product of innovative scientific activity,"[1] wrote George Basalla, "is most likely a written statement, the scientific paper, announcing an experimental finding or a new theoretical position (Basalla 1988, p. 30)." However, normal science thrives on the same additive dynamic as does technological innovation and often a breakthrough represents no more a conceptual leap than the results of an "ordinary" experiment. Only the context is different. A publication is no more the "end product" of science than the iPhone is the end product of cell phone research (although they may misleadingly appear to be more than they are (Rudwick 1985, pp. 434–435)). Fritz Haber's 1905 publication on ammonia synthesis, if this were considered to be "the paper," fits into a continuous scientific narrative on the development of classical thermodynamics and its transition to a basis of statistical and quantum mechanics. Energy science in the mid-1800s led to an understanding of heat and work which was applied to gases; this knowledge was adapted to chemical solutions and gave birth to physical chemistry. Subsequent steps came in the form of Walther Nernst's third law with experimental confirmation by Fritz Haber. The closure of classical thermodynamics then provided an avenue for the application of more modern physical theories (statistical mechanics and quantum theory) to these same systems. Each of the achievements was marked by one or more publications but did not necessarily result in a precipitous jump in quantity or quality of physical knowledge. While each achievement resulted in improved abilities of prediction (and even clairvoyance), they also suffered from inaccuracy and confusion.

---

[1] The terms *novel* or *progressive* seem more appropriate here than *innovative*. For the same reason, I use the term *breakthrough* or *discovery* in a normal-scientific context rather than *innovation*.

Scientific research has also been treated analogously to technological innovation in that different, smoothly varying outcomes are assumed possible (Rudwick 1985, p. 450). In what is, incidentally, a comparatively broad discussion of scientific progress, Kline and Rosenberg discussed Galileo's contribution to mechanics (Kline and Rosenberg 1986). "Without the telescope," they wrote, "we would not have the work of Galileo, and without that work we would not have modern astronomy and cosmology [. . .] It is probable also that without Galileo's work we would not have had what we now call elementary mechanics until a much later date, *and perhaps not at all* [italics added]."

Here again, the idea of the heroic inventor in technological innovation is invoked, or the heroic scientist, the general centrality or inextricability of whom I have argued against because it is so incompatible with continuous progress. While they do appear, sometimes to great effect, the success of a single person is often due to (ideal or less than ideal) circumstances outside their control or they are later made into something they never were (Fleck 1980, p. 61), (Holdermann 1960, pp. 95, 96), (Kasperson 1978; Simonton 1984), (Becker 1984, pp. 300–301), (Rudwick 1985, p. xxii, 15, chapter 2, pp. 411–428), (Basalla 1988, pp. 57–66), (Padgett and Ansell 1993), (Csikszentmihályi 1996, pp. 330–336), (Adichie 2009). One reason given is nationalism, which is certainly true in the case of Fritz Haber. Another is the confusion when continuous technological change leads to outsized social or economic consequences (the "catalytic effect") (Rudwick 1985, p. 438), (Barley 1990). Again, this is true with Haber. His scientific contribution was not outsized and, in fact, Nernst's theoretical solution held greater scientific consequence. But Haber's was the final step before the power of physical chemistry could be unleashed in the form of the high pressure catalytic chemical industry. The circumstances shaping the arena for discovery were decisive.

Returning to Galileo, did Kline mean to say it is *probable* that if it were not for Galileo, we would now have an *entirely new* theory of mechanics? I think we would not. There is again the argument against the hero. "Galileo gave to modern science the quantitative experimental method," wrote Hans Reichenbach. "Yet this general turn toward experimental method can scarcely be regarded as the effect of one man's work. It is better explained as the result of a change in social conditions. . .[that] led naturally to an empirical science (Reichenbach 1954, pp. 89–99)." It is certainly possible that without Galileo the development of mechanics would have been set back, or, with a substantially smaller probability, that we would have no theory at all. Where would that leave us after having entered a planetary reality of "an infinite number of mutually separated and isolated, hard and unchangeable—but not identical—particles (Koyré 1968)"? Truly with no theory of mechanics for all times? Would we eventually develop an altogether new theory? We currently have three options to choose from: "classical" Newtonian mechanics, quantum mechanics, and relativistic mechanics. While it is clear to us today that the first is an approximation that works at the dimensions and speeds of the reality occupied by the human being, it is a very good approximation and is substantiated by the other two theories if appropriate constraints are applied. We still rely on classical mechanics for many calculations—at our order of magnitude there is still no better or simpler option.

And no one expects to find one. Considering our place in the universe, it is unlikely quantum mechanics or relativistic mechanics would have been developed before classical mechanics as the quantum of action and the invariability of the speed of light are needed for these theories. Likewise, it is improbable that observations made by human-like beings working under the assumption that the universe exists in a state similar to the one with which they perceive to directly interact would have led them to a description wholly different than classical mechanics. Velocity or momentum may have been used as basic units instead of distance and time, but the underlying physics of the interaction of solid objects moving in the way we humans can most easily observe and contemplate would have remained the same. Having said this, if we were the size of an electron[2] we may first have had a functioning quantum mechanics (which we would know simply as "mechanics") while current scientists may be slogging away on a theory of "continuous mechanics" to describe the mysterious world of huge conglomerations of matter on the unfathomable order of magnitude of kilograms. Would our observations have already reached the scale on which relativistic mechanics is needed? We find ourselves at dimensions between the domains of the quantum and the astronomical. Both realms have become accessible to us, making it a unique vantage point. Or is it? Perhaps it is a normal step beyond the scales that one inhabits to understand how to observe the scales above and below and we are still not able to see past the initial step in each direction. The paradigm we use depends not only on how we choose to see the universe, but how we are able to see the universe; the theory or paradigm that results from research is dependent on the way the universe (or system) appears to be as it is developed. Without the work of Galileo himself, therefore, I think we would still be in control of a classical theory of mechanics because of the kind of instinctively "natural" universe it describes to us. In contrast, quantum mechanics is almost a century old and it has still not been grasped by most physicists. Could such an absurd view of the universe really precede classical mechanics?

If we were able to peer further or observe more, we might discover a form of contemplation different from anything we already know, but for each perspective there is only one paradigm that fits. By interpreting an observation in a certain way, a useful theory may be built upon it, though a different interpretation may show that theory to be limited, as surely all theories are (or will eventually be shown to be) with respect to some perspective on the natural world. No one contends that any single paradigm they employ describes all aspects of the universe in a perfect manner. But a useful paradigm contains representations of the truth and a set of rules that allows for communication and continuity in research, even after a new paradigm achieves greater precision or more accurate descriptions. "Ptolemy's astronomical system," wrote Reichenbach, "also called the geocentric system, is still used today to answer all those astronomical questions which refer merely to the aspect of the stars as seen from the earth, in particular, questions of navigation. This practical applicability shows that there was a large measure of truth in Ptolemy's system."

---

[2] This is meant as a thought experiment.

The only drawback was "the imperfect state of the science of mechanics at the time (Reichenbach 1954, pp. 97–98)." Even though we now have a more mature theory of mechanics, it is conceivable to generate new knowledge within Ptolemy's outdated paradigm and create new technology, perhaps a consumer navigation device for the hobby sailor. Physical chemistry, as Haber employed it, was also very useful, even in its classical form; later it was extended to reveal more about the behavior of nature. Its first incarnation is, however, not illegitimate.

The kind of paradigm or theory which functions correctly (however naive or incomplete it may be) must be based in some way on the "spirit" of science. There must be some observation and an effort at interpretation using physically insightful and meaningful tools that deepen the meaning of the observation. That the sun rises and travels across the sky is a meaningful observation, for example. That the greek god Helios drives his chariot each day toward Oceanus is not a meaningful interpretation; it does nothing to explain or embed itself in the further workings of the universe, except, of course, that our world is inhabited and manipulated by various gods, specialized in their own self-absorbed rituals. This "fact," however, will never be substantiated by any observation because it is not true. There are also aspects of the scientific perspective of the universe, including energy conservation and the classical force of gravitation, that cannot be derived from more integral, basic ideas. However, they are based on repeated observation, give us insight into how the universe may function or behave, and allow us to generalize our observations to derive further useful knowledge (while keeping in mind that these are stated "laws" and not proven fact). As long as a theory operates in this way, it can be considered science and one may say that it contains a "measure of truth." That is why, scientifically, Helios is no longer of any use to us.

Later, the role of different paradigms will be considered further, but first we return to where we began: normal science within a paradigm, which, in contrast to technology, is adorned with objectivity. The application of electromagnetism to the electric motor, for example, did not predetermine its final configuration (Basalla 1988, p. 43). The view of electromagnetism of mutually-inducing electric and magnetic fields did, however, lock in Maxwell's equations. Every subsequent derivative description followed from there.

When considering science, the objectivity within a given framework is a commonly misunderstood characteristic. Science is often viewed by the general public as thoroughly objective and dealing in hard facts; within other expert or academic circles, the same scientific activity is often considered, at some level, to share the subjectivity found in technological innovation or progress and change in medicine, politics, art, sports, business, and other fields. It is neither of these. Normal science is subject to strict rules. The way in which the rules must be followed, however, is limited only by creativity and ingenuity. Like many endeavors, it is an art form. The goal of the scientist, be it the creation of knowledge for the common good, the pleasurable or perhaps selfless "need to know," or the achievement of lasting acclaim, may seem only fleetingly definable. "I admit," wrote Subrahmanyan Chandrasekhar, in trying to capture the essence of science, "that these are things which cannot be defined any more than beauty in art...(Chandrasekhar 1987, p. 13

and chapter 4)." Add to this nebulousness the challenges of developing a new physical theory and the complexities of science seem to escape the simple two-part description consisting of normal science and paradigm shifts.

In describing the Haze further, I focus on three factors that illustrate science and scientific discovery in a way that reveals the overlap and interdependence of fact-gathering and paradigm shifts and how a scientific breakthrough is the result of this relationship. In doing so, I maintain the definition of "science" as the interpretation of experimental results, obtained in a transparent and reproducible way,[3] within a theoretical framework. The three factors are:

(1) The existence of (what behaves like) an objective truth—a fact or optimized set of system conditions—within a given paradigm. This objectivity results from the rigorously defined way a particular aspect of nature[4] is perceived and treated by researchers bound to that paradigm. The word "subjective" refers to activity not fully within this context.

(2) The environment of knowledge exchange within which a scientist works includes living colleagues but also the paradigm—a body of knowledge built up over generations—which provides a framework in which they act. Whether a scientist is interacting with his colleagues (social interactions) or facilitating an exchange between established theory and experiment (epistemic exchange), the *successful* transfer of scientific knowledge depends on the objectivity in 1).[5]

(3) The value of scientific results, concepts, equations or laboratory procedures is not immediately or completely identifiable, nor is there a general chain of development by which value is increased. Rather inherent value is recognized or assessed.

I will not attempt to elaborate on the value of science here, but a few words will help explain my use of the term.

Scientific value derives from an advance that allows or forces us to view a system in a new or more fundamental way (Siggelkow 2007)—that is, *it makes us think differently* (Fortus 2018). The only immediate limitation is our creativity when combining theory and experiment. The inability to think outside the box has proven to be a real problem. As Gerhardt Ertl put it while contemplating the cleverness of

---

[3] I do not include the process of measurement in this definition—only the knowledge of the conditions under which the measurement was made. I refer to science in a broader sense using the term "scientific endeavor."

[4] For example the *Fixum* or *passives Wissenselement* from Fleck (1980, pp. 68–70, 109–111, 125–133), the "token of reality" from Polanyi (1962, p. 132), "the rules of the game" discussed by Feynman (1979), or the rules, structures, and symbols described by Csikszentmihályi (1996, pp. 36–41).

[5] In the words of Ernest Rutherford (1938, p. 74): "Scientists are not dependent on the ideas of a single man, but on the combined wisdom of thousands of men, all thinking the same problem, and each doing his little bit to add to the great structure of knowledge which is gradually being erected."

the van't Hoff Box in Appendix A (Ertl 2018): "We have the money—what we need now are ideas!"

Polanyi identified three contributions to value in science: certainty (accuracy), systematic relevance (profundity), and intrinsic interest (Polanyi 1962, pp. 135–136). The first two can be grouped together into what I call "usefulness" with identifiable value within a paradigm. The third, a subjective factor, plays a role in the identification of pertinent scientific questions along with "our vision of reality, to which our sense of scientific beauty responds." However, while intrinsic interest may make something fascinating, it does not make it scientifically valuable.

I find the best way to express the value in science is as the integral of all usefulness over time (which may be positive, negative, or zero):

$$Scientific\ Value = \int_{t=\tau_o}^{Present} Usefulness \cdot dt$$

All that remains is to define the time-dependent function "$Usefulness$." It will not be a straightforward task; we may just have to get creative. One widely considered possibility that must be discounted stems from the detrimental misconception that productivity and proliferation alone correlate with usefulness and scientific value. They do not—except perhaps inversely.

Returning to the three factors above, I again stress they are not meant to suggest the primacy of a scientific discovery over a technological innovation. While there are many innovations which have occurred seemingly without direct input from basic scientific research, it would be erroneous to declare any modern scientific discovery came about without the aid of technology. I am also not suggesting the superiority of the scientific method over approaches that have been successfully employed in other fields. Rather, I am stating that the scientific method has enabled significant progress in some areas. Here, the focus is on the type of scientific discovery that could be viewed as a *decisive event* or "an especially important event that provides a major and essential impetus to the innovation. It often occurs at the convergence of several streams of activity. In judging an event to be decisive, one should be convinced that, without it, the innovation would not have occurred or would have been seriously delayed (Globe et al. 1973, p. 2)." Such a decisive event need not be a scientific discovery, but it certainly may be. Conversely, a significant scientific discovery need not lead to technological innovation. This situation, in the context of ammonia synthesis, represents the most inclusive process chain leading to innovation, though the presence or absence of technological consequences has no effect on the dynamic of the Haze itself. Prior to the arrival of the decisive event (in this case, the scientific breakthrough) there are a series of occurrences which, to varying degrees, may be drawn into a relationship to the discovery. These may be classified as *nonmission-oriented research* (NMOR), or "research carried on for the purpose of acquiring new knowledge, according to the conceptual structure of the subject or the interests of the scientist, without concern for…application" and *mission-oriented research* MOR, or "research carried on for the purpose of acquiring new

knowledge expected to be useful in some application (Globe et al. 1973, p. 3)."
It is this change from overwhelmingly NMOR to MOR with respect to a specific
breakthrough with is embodied by the confluence of factors. It is this dynamic
which illustrates the *condensation* of the Haze into the arena for discovery and
the *precipitation* of the breakthrough.[6] It is this arena that separates the preceding
"normal" fact-gathering from a subsequent scientific discovery. The individual steps
of scientific advancement may appear identical in both cases with only the context
supplying a means of differentiation. It is not a linear convergence because feedback
mechanisms interlink aspects of basic science and technological progress. During
the condensation, uncertainty and differences of opinion among the practitioners
of a discipline concerning the exact knowledge defining a breakthrough are vastly
reduced, the spread of choices narrowed (Fleck 1980, pp. 15–16, 72–73, 110),
(Rogers 1995, pp. 13, 36–37), (Rudwick 1985, p. 455), (Kline and Rosenberg
1986; Padgett and Ansell 1993). Available information allows the astute observer to
identify which experiments (and which aspects of those experiments) are necessary
to articulate the paradigm in the way that reveals the breakthrough, as if bringing
closure to the acute questions that first exposed the arena for discovery. Let us
consider the events leading to ammonia synthesis from the elements in Fig. 17.1.
There are certainly external factors that shape the trajectory of a discovery, such
as the Margulies brothers offering Haber financial compensation for his research or
the academic political shuffle within which Haber and Nernst were embroiled. Here,
however, we restrict ourselves to scientific developments.

The transition from NMOR to MOR is particularly clear in this case, as is the
convergent nature of the NMOR or MOR on its own. Consider, for example, Haber's
initial results in 1905 compared to the increased accuracy that followed (as well as
his reinterpretation of the initial results in terms of the increased knowledge base
described in Part II, Chap. 11). The accumulation of MORs near the discovery—
not necessarily in a temporal, but rather in a knowledge-based sense—may be
a consequence of the difficulty of actually achieving the breakthrough, in other
words, of properly embedding the last step in an attempted paradigm articulation
within the complex network of knowledge that preceded and led to it (Klein 2016a).
"It happens again and again," wrote Wilhelm Ostwald in 1908 (Ostwald 1908,
pp. 22–24),

> ...that of all things, the very last step, with which the new thought would be completely
> perfected and able to stand in opposition to old ways of thinking, is usually forgotten,
> overlooked, or disregarded by the creator of the new idea.[7]

While Ostwald may have been referring to groundbreaking "new thought" more
in line with paradigm change, the travails of normal science will not have been

---

[6] So is revealed my reliance on terms borrowed from physical processes.

[7] "Es zeigt sich immer wieder...daß gerade der aller letzte Schritt, durch welchen der neue
Gedanke zur vollständigen Abrundung und zum vollständigen Gegensatz gegen den alten
Gedanken gelangen würde, von dem Schöpfer dieses neuen Gedankens meist vergessen, übersehen
oder vernachlässigt wird."

Event	Approximate Date	Relevance to Ammonia Synthesis	Research Type with Respect to Ammonia
Identification of inorganic elements and compounds	~1800	Nitrogen, hydrogen and ammonia exist	NMOR
Analysis of plant and animal tissue	~1800	Ammonia exists in these tissues	NMOR
Identification of organic compounds	1800-1840	None	NMOR
Research stations and mineral theory of plant nutrition	1820-1850	Plants assimilate nitrogen in inorganic form (and not from humus)	NMOR
Plants assimilate nitrogen in fixed form	1820-1850	Need for source of fixed nitrogen	MOR
Initial work on energy science	1830-1850	Basis for thermodynamics and physical chemistry	NMOR
Organic compounds useful for dye synthesis	1860s	None	NMOR
Establishment of thermodynamics and physical chemistry	1860-1890	Enables calculation of ammonia equilibrium	NMOR/MOR
Establishment of infrastructure, capital and know-how in the chemical industry	1860-1900	Enables experimental work on ammonia synthesis	MOR
Attempts by Ramsey and Young, Le Chatelier and others to establish ammonia equilibrium	1880-1900	Scientific pursuit of ammonia synthesis	MOR
Ostwald's attempts at ammonia synthesis	1900	First combination of all necessary elements for ammonia synthesis	MOR
Breakthrough by Fritz Haber, Walter Nernst and others	1903-1908	Arena for discovery	MOR

Fig. 17.1 The scientific developments during the confluence of factors leading to ammonia synthesis and their relationship and relevance to work between 1903 and 1908 by Fritz Haber, Walther Nernst, and others. NMOR stands for *nonmission-oriented research*, MOR for *mission-oriented research*. See text for details

far from his mind: just a few years earlier, he had concluded his failed attempt at ammonia synthesis. After Carl Bosch reported his inability to reproduce Ostwald's experiments at BASF in 1900 (Part II, Chap. 9), the latter was "so exhausted that I could no longer bear further engagement with these things (Ostwald 1903, p. 287)."[8] Continuing with this line of thought, again in 1908, Ostwald commented that the aforementioned "creator" had "no more remaining strength" to finish his task.[9] He had been close to his goal and had defined the steps Haber would later successfully implement. It was only the attention to one detail, that minuscule and reproducible quantities of ammonia should result from his experiments, that evaded Ostwald's consideration. Had he recognized this experimental possibility, he would have fully embedded his results within the framework of physical chemistry. In fact, the other researchers investigating ammonia synthesis at the end of the nineteenth century suffered similar fates. Ramsey and Young also measured values that were too large, whereas Le Chatelier lacked the necessary experimental ability when working with high pressures. Edgar Perman was an exception; in disregarding the principles of physical chemistry, he tried to articulate the wrong paradigm altogether. All of them were working in a time when the complete set of puzzle pieces was available. They simply had to be assembled in the right way—a task that can be as formidable as identifying the pieces in the first place. They must be assembled perfectly, of course, because there is no workaround. The objectivity of science places strict constraints on the acceptability of the final answer.

It is certainly not a general property of the Haze that it exhibit this level of simplicity, which again shows why ammonia synthesis is such an informative example of scientific discovery. In a general model of scientific progress, there must be the allowance for greater oscillation between NMOR and MOR and less demarcation of discrete, independent classifications of science so that any generic form of the Haze remains arbitrary. Certainly traditional, cultural, societal, political, or any number of external factors (including luck) also influence the speed at which the Haze condenses. They do not, however, have any influence on what precipitates.[10]

During the transition period, there is a recognition of scientific value when the NMOR finally gives way to MOR and results become consciously applicable to further research. At this point, the position and relevance of the knowledge within the paradigm becomes apparent as does its usefulness, although a specific determination of value is only in regards to the corresponding discovery. Any NMOR, at an earlier date, may have given way to MOR (with respect to another subject) and any new MOR (with respect to ammonia synthesis) may already be considered MOR with respect to another application. Again, we are confronted by the delineation of events and can see why a schematic approach to one discovery

---

[8] "...so erschöpft, daß ich eine weitere Beschäftigung mit diesen Dingen nicht mehr ertragen konnte."

[9] "...keine Kraft mehr übrig..."

[10] The word "congeal" is also fitting (Rudwick 1985, p. 418).

helps isolate the underlying dynamic. This approach of assessing the value of science is abstracter than outcome-based methods (for example, assigning greater value to research that results in patents), but it is more fitting with respect to basic research (Rogers 1995, p. 135).

The procession toward scientific discovery can begin without any "need" for a solution. Later, after the MOR has evolved out of the NMOR and has made a principle contribution to the creation of the arena for discovery and proper insight has led to the breakthrough, revision is no longer required. Rather the revision comes just before the breakthrough as the MOR "circles" the correct answer and the scientists are aided not just by knowledge and expertise but also by intuition. This development is unlike technological innovation, where the step from invention to innovation is marked by the identification of a need and often requires a process of revision before the resulting product is market-ready (Basalla 1988, p. 23). Interestingly, the reason why a new technology (invention) may come to be seen as an innovation can be due to the existence of the technology itself (Langrish 2017; Sgourev 2015). That is, the perceived need for an innovation can either precede or follow the invention leading to it. In a scientific breakthrough, the arena for discovery where a distinct and definite problem has been revealed (to which a solution surely exists), having followed from the confluence of factors, must precede the successful breakthrough. The same is true even if the discovery is not recognized at the time the original work is completed; in such a scenario, the arena is not yet mature and the full articulation of the paradigm and resulting realization that a breakthrough has occurred is not yet possible. The arena does not identify need or market potential, it is the assemblage of knowledge and experience that reveals a relationship in nature according to the paradigm. From a scientific perspective, one particular result is not "needed" more than any other. The last puzzle piece is often crucial because it is the last, not because it is the most important. In this way, scientific results are not the same as technology—if a scientific result has value, it will eventually be identified through the mere circumstance that we continue to perform scientific work. Its place within the existing knowledge will be found. This activity inherently pushes us toward appropriate questions and, given enough time, the answers to those questions (Polanyi 1962, p. 135). It is as if something were nudging our understanding in a direction that is clear only in hindsight while the answer is still hidden in the paradigm. Can we represent this vague notion more concretely?

Our path through the research on ammonia synthesis would have gained considerable complexity with increased reference to personal relationships or local and international politics; this information would have added further insight and contextualization to the science, but it would have detracted from the opportunity to exploit the "relative simplicity" of the discovery to provide a simplified view of the otherwise complex structure of the Haze. Focusing on the science allows us, for example, to derive the schematic diagram in Fig. 17.2.

The condensation of the Haze may be represented with a cone whose axis lies parallel to the direction of time, growing denser as it narrows. The point of the cone marks the beginning of the arena for discovery in which the breakthrough

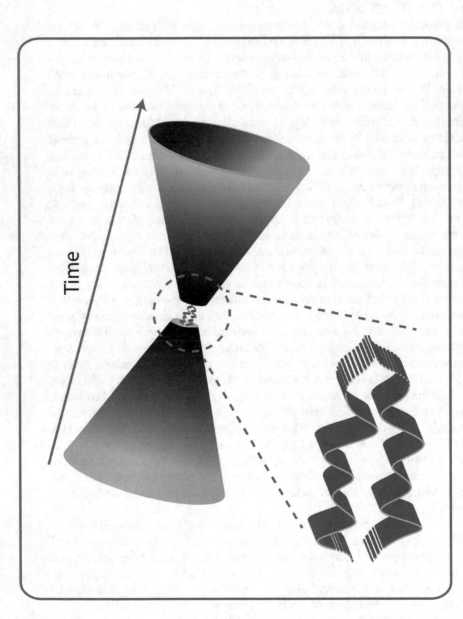

Fig. 17.2 Schematic diagram of *The Haze*. The condensing cone on the left represents the confluence of factors leading to an arena for discovery in which the breakthrough, a unique and irregular structure, is precipitated. These together comprise the Haze. The right-hand expanding cone represents the consequences of the discovery, in particular the technological change that becomes ever-more interlinked with elements of our lives. The inset shows a close up of the breakthrough and its intricate structure. Image by Birgit Deckers at the Max Planck Institute for Chemical Energy Conversion

is precipitated. In general, it is an episode drawn out in time (not a single event) with a unique, irregular structure as illustrated by the events of Part II. If the scientific breakthrough is a *decisive event* for technological innovation, as it was with ammonia, the subsequent developments may be represented with a second cone also lying parallel to the time axis; its point is at the end of the breakthrough and extends away from this moment, becoming dilute. The expansion contains technological developments that permeate and reconfigure our daily lives (Padgett and McLean 2006). It represents a broadening effect over time, not necessarily a "catalytic," non-linear reaction, and may include desirable or undesirable, direct or indirect, as well as anticipated or unanticipated consequences (Rogers 1995, pp. 30–31), (Sveiby 2017). In the case of ammonia synthesis, some of these are the production of industrial fertilizer, explosives, pesticides, and poison gas, as well as the high pressure catalytic industry. The expanding cone also contains further scientific progress. New ideas amend or replace older ones, such as the application of statistical mechanics or quantum theory to thermodynamics (Kuhn 1970, p. 21), (Fleck 1980, p. 29). This type of expansion of a novel concept or device is not limited to science; it is seen across the spectrum of human activities (Brooks 1973; Bush 1945a; Sgourev 2013, 2015).

The use of a cone is indeed a stark simplification of the actual structure of the Haze, but its solidity reminds us that holes or jumps in advancement are rare; the fluidity is a consequence of the continuous nature of scientific and technological advancement. The difficulty is, rather, the way in which we choose to isolate events that are at once connected but also in some way clearly separate. It is a problem known in defining different classes of technologies or, more generally, different artifacts of many kinds (Murmann and Frenken 2006; Vedres and Stark 2010).

The double-cone structure in Fig. 17.2 may appear as a linear approach to a non-linear dynamic but only because the complexity has been hidden away for ease of management. The trick of simplicity to achieve transparency is an old one:

> Every concrete process of development finally rests upon preceding developments. But in order to see the essence of the thing clearly, we shall abstract from this and allow the development to arise out of a position without development (Schumpeter 2012, p. 64).[11]

Put another way: "the main direction of the development, taken as an idealized average, would have to be drawn separately and at the same time [as all other lines of development] (Fleck 1979, p. 15)."[12]

An objection to this schematic diagram is the neglect of influence of prior or external science and technology on the condensation of the Haze. To represent this interaction we can draw as many overlapping structures as needed in an ever-widening array (Fig. 17.3). Some may be double cones while others are single

---

[11] "Jeder konkrete Entwicklungsvorgang endlich beruht auf vorhergehenden Entwicklungen. Um aber das Wesen der Sache ganz scharf zu sehen, wollen wir davon abstrahiern und die Entwicklung sich aus einem entwicklungslosen Zustand erheben lassen (Schumpeter 1952, p. 98)."

[12] "...müßte man die Hauptrichtung der Entwicklung, die eine idealisierte Durschschnittslinie ist, gleichzeitig separat [von den anderen Entwicklungslinien] zeichnen (Fleck 1980, p. 23)."

Fig. 17.3 Illustration of the interdependent nature of scientific discovery and technological innovation based on the schematic diagram of the Haze in Fig. 17.2. Three structures are present: the double-cone, the condensing cone (the Haze) representing a scientific breakthrough not immediately leading to a new technology, and the expanding cone representing a new technology not dependent on a scientific breakthrough. In reality, the density of shapes is much greater

structures consisting of only the Haze (a scientific discovery not immediately leading to a new technology), or an expanding cone (a new technology not dependent on a scientific discovery). A complete picture would consist of an unwieldy number of cone structures extending the length of any relevant time axis. This notion would not help in understanding the underlying nature of a single breakthrough, but it elucidates the sophisticated interaction between science and technology as well as the problem of delineation: a stand-alone object that also exhibits a strong dependence on scientific, technological, and other events.

Later, one addition to the schematic double-cone diagram will be discussed: a transitional stage between the breakthrough and the expanding second cone, representing hybridized scientific and technological work (Fig. 19.1).

Chapter 18
Dynamic of The Haze

Here we turn to mechanisms of knowledge transfer in the Haze, initially within a paradigm before expanding to cross-paradigm exchanges. Before we do, the reader is reminded of one aspect of the following analysis. The use of theoretical tools to describe the interactions which lead scientists to new ideas or combinations of knowledge can make them appear routine, even as arranged occurrences. This strategy is helpful in reducing complexity to a manageable level. However, it also obscures the random and stochastic nature of these interactions. In general, it is not possible to plan them and often it is not evident until after the fact that such an exchange "triggered" the recognition of a new connection (Graßhoff 2008). This obscurity is the cause of the insurmountable complexity of the Haze; it is dependent on what we, at this point, can only classify as accidental occurrences. The more experience we gather with scientific breakthroughs, the better we can generalize them, however, they remain unique incidents resulting from a vast number of interdependent events. The story of one breakthrough holds hints about how another breakthrough will unfold, but cannot explain it fully. The physical and mathematical understanding of how and why the Haze condenses in one particular situation is contained in the breakthrough itself; the ability to understand how the pieces fit together requires the breakthrough to already have happened. If we wish to influence these processes, however, we may be able to *increase the probability* of such beneficial circumstances.

The narrative in Parts I and II illustrated several interactions between Haber and Nernst (and the resulting scientific progress) with only limited attention to the social setting. In the case of ammonia synthesis, the setting was simple: a small number of actors (or alters), manageable geographical distances, and adequate avenues of communication. In consideration of the development of scientific thought, it has been informative to consider the breakthrough without a theoretical social aspect (Rudwick 1985, pp. 411–435). Many scientific breakthroughs occur in arenas for discovery of limited social scope, making this perspective a common one. Can a reliance on case studies, or anecdotes, lead to conclusions similar to those of

© The Author(s) 2022
B. Johnson, *Making Ammonia*, https://doi.org/10.1007/978-3-030-85532-1_18

studies involving statistically significant numbers of actors encompassing an array of experiences (Lalli et al. 2021)?

Successful scientific advances are a result of direct epistemic transfers of information between theory and experiment or from social interactions which achieve the same outcome (Rudwick 1985, pp. 6–15). However, scientists operate in an environment that also allows for social interactions which are not of this type. What are the consequences of the different kinds of exchanges? Considering the nature of a scientist's social network, we include living adherents to the accepted body of theoretical and experimental knowledge making up the content of the paradigm.[1] These are members of the scientific social setting who are able to personally engage in meaningful transfers of information. In order to incorporate the paradigm itself into the social setting, we can also include deceased colleagues who have made contributions to the body of knowledge (Rutherford 1938, p. 74), (Fangerau 2010, p. 131). Transfers facilitated by a scientist between dead colleagues (directly between theory and experiment) are epistemic transfers. Examples from ammonia synthesis include the way in which Nernst contrived and applied his version of the third law or how Haber was able to combine the empirical studies of the catalyst with the strict physicochemical equilibrium determination. Social interactions between living colleagues, however, can have varying outcomes. An example of a non-epistemic knowledge transfer is Nernst's behavior at the meeting of the Bunsen Society in Hamburg (Part II, Sect. 11.1.4)

For a closer examination, let us consider some social network theory applied to the case study of ammonia synthesis. These theoretical tools will help us differentiate between specific types of knowledge transfer to sharpen the boundaries of what is and what is not part of science—even in the limit of small groups of actors. As we will see, some elements of the macrodynamics associated with the scientific endeavor do not result in the identification of any new scientific value because they are not part of science at all.

In scientific settings, potentially creative isolates may be at work at the edge of their social networks. (The influence of social settings has been linked to innovation and creativity, stemming particularly from loosely bound adherents to a group (isolates) (Ibarra and Andrews 1993; Perry-Smith and Shalley 2003).) However, what amounts to "social isolation" in science does not only apply to one's current social setting. It also applies to the paradigm. Were he talented enough and supplied with proper knowledge of physical chemistry and catalysis, Fritz Haber could have completed his scientific experiments on ammonia synthesis in complete physical isolation. He could have worked in a cave without access to experience or external input from living colleagues. In other words, to achieve his breakthrough, it was possible for Haber to rely purely on information available before he began his work. He needed an apparatus, but this is the realm of engineering (in Part II, the contributions of his technical assistants are outlined), which allows only the

---

[1] Today scientists are often adherents to multiple paradigms spanning the investigation of different components of complex systems (see Chap. 21).

production of experimental data. As it turned out, the interaction with Walther Nernst at the meeting of the Bunsen Society in Hamburg was not just accelerative (as it would have been in the idealized case of a "discovery in a cave"), it was needed to extend the limits of both Haber's and Nernst's understanding of physical chemistry as neither was in command of the complete set of physicochemical knowledge available at the beginning of the research. "*Neither* of the initial alternatives," writes Rudwick in the case of the Devonian Controversy, "...had a monopoly in the requirements for victory...(Rudwick 1985, p. 405)." An incomplete knowledge base is often the case and scientists require interaction with one another, or the chance to capitalize on "the utility of people," in order to successfully incorporate creative elements into their work (Becker 1984; Kanter 1988; Kasperson 1978), (Csikszentmihályi 1996, chapter 6, pp. 294–296), (Obstfeld 2017; Perry-Smith and Shalley 2003). The reliance on actual human exchanges illustrates the complexity of the real-life social setting of a scientist (Rudwick 1985, p. 456). In practical situations, the mere transfer of concepts between theory and experiment is unlikely to be successful on its own, because one person will not be in possession of a complete set of knowledge.

In real scientific work, there is much trial and error, conjecture, and opining about what the right answer might be. New ideas are continuously advocated and then accepted or rejected, both with and without apt consideration (Stigler 1982). These decisions are based on two strategies of cooperation. One is that "the certain path to feeling creative is to find a constituency more ignorant than you and poised to benefit from your idea (Burt 2004)." Burt, I think, was making a theoretical point about the power of bridging a "structural hole (Burt 1992, 2004)." A different dynamic can also play out. I once heard a senior colleague say something to the effect of: the way to be successful in science is to collaborate with people who are smarter than you and learn from them. It is the same dynamic of knowledge transfer but depends on who is giving and who is receiving—it is easy to see which party will tend to be more critical and which more deferential. Eventually, everyone will have their turn at each end. However, the decision-making process has no effect on what the correct answer actually is.

What, then, is the mechanism of transfer?

Borrowing the term from social network theory for the purposes of our examination, the answer is *brokerage*. Put simply, the concept of brokerage refers to behavior that links individuals from different groups via the closure of an open triad (in other words, it bridges a structural hole). Brokerage is also possible through the act of bringing individuals together who know each other but had not yet thought to collaborate in the manner proposed or enabled by the broker. Such activity may presuppose the existence of special (social) skills, self-motivation, the ability to motivate others, or the establishment of trust (Burt 2004; Fligstein 2001; Obstfeld et al. 2014; Padgett and Ansell 1993), (Obstfeld 2017, chapter 1), (Sgourev 2013, 2015). It also provides a connection between micro- and macrodynamics (Sgourev 2015), (Rudwick 1985, p. 14). The interesting cases are those acts of brokerage that result in combinations of knowledge leading to something new (a thing or idea), possibly triggering a non-linear effect. Strictly speaking, brokerage is a social interaction with many outcomes, some of which could have the effect of an epistemic transfer. Successful scientific advancement depends on the linking of theory and experiment (epistemic transfer) by means of a novel pathway along with relevant social interactions that amend gaps in an individual's knowledge (Rudwick 1985, p. 405).

There are three kinds of brokerage. The most basic is conduit brokerage in which "the broker provides value to one group by providing them with needed resources derived from another group. The potential for providing value through conduit brokerage is a function of the differences between the parties connected by the broker (Obstfeld et al. 2014)." The more valuable the information, or token, the higher the syntactic boundaries (common lexicon) and semantic boundaries (requires translation) become (Carlile 2004), (Obstfeld 2017, pp. 33–34). The next type of brokerage, *tertius iungens*, or "the third who joins," includes knowledge transfer but goes beyond conduit activity and "is most opportune when the broker detects opportunities to connect complementary, rather than redundant, alter [individual] attributes such as resources and abilities. At the same time, *iungens* brokerage connecting those with differing ties or attributes brings with it the corresponding challenge of coordinating dissimilar backgrounds and interests (Obstfeld et al. 2014)." However, successful knowledge transformation can lead groups or individuals to change their position and embrace new and innovative strategies (Obstfeld 2017, p. 56). These two types of exchanges can facilitate an epistemic transfer and drive the dynamic that advances science. While *tertius iungens* brokerage harbors the potential to combine knowledge into substantial advances, there is also the risk of the "action problem." This complication arises when useful combinations of knowledge are possessed by groups which are uncoordinated and, therefore, cannot act in concert. The heterogeneity promises greater potential benefits for those involved, but also a higher risk of failure (Obstfeld et al. 2014). The inverse of the action problem is the "idea problem," in which dense networks have the ability to react but may lack the ideas with which they can produce novel value (Granovetter 1973; Obstfeld 2005). This situation is in play in any interdisciplinary pursuit without brokerage. Science itself would function for some time without significant knowledge transfer, but eventually conduit or *tertius iungens* activity is required because no one scientist can independently achieve all necessary epistemic transfers.

There is a also a third type of brokerage called *tertius gaudens* (Burt 1992), or "the third who rejoices (when the other two parties are in conflict)." "*Tertius gaudens* strategies involve the restriction of alter-alter activity by either keeping certain alters apart or actively cultivating alter-alter tension in a given interaction...(Obstfeld et al. 2014)." The broker employing this practice "profits by maintaining separation between alters [players, individuals]...(Obstfeld 2005)." This type of social interaction, which is also present in a scientist's social environment, is not and does not facilitate an epistemic transfer and is without permanent effect on scientific progress (the interpretation of experimental results in the context of a theory). It is, however, found in two endeavors that are tightly intertwined with the social dynamic of science: engineering and politics (in an academic setting or with respect to funding decisions). Both these pursuits have subjective aspects so that *tertius gaudens* microdynamics, whether intentional or not, can alter the perceived best and, therefore, chosen outcome. In science, the result leads to social artifacts such as research strategies, selection of methods, accepted solutions, and funding decisions.

In the conclusion of Part II, the importance of "microcomplexities" between 1903 and 1908 were discussed as they emerged within the arena for discovery of ammonia synthesis. There is a gap between microprocesses and observed macro-scale characteristics which may be closed by appropriate aspects of brokerage (Kanter 1988; Obstfeld 2005; Sgourev 2015). Due to this gap, science can appear from the macro-level to contain subjective social elements resulting from political dynamics and engineering decisions. That is, science appears to behave like any other field (engineering, politics, art, business, sports, etc.), especially if *tertius gaudens* microdynamics are considered. At the micro-level, however, *tertius gaudens* (and some conduit and *tertius iungens*) brokerage activity does not result in an epistemic transfer and makes no contribution to scientific advancement. The solution to a scientific investigation is determined by the paradigm and it is only a question of time until the twists and turns of normal science (perhaps created by various social interactions or erroneous conclusions or results) lead to the correct answer. The solution itself, and the value of this solution, is path-independent because a theory is not impacted by our wants and needs (Polanyi 1962, p. 4). The only permanent result of *tertius gaudens* brokerage in changing a scientific outcome is to delay it for so long that a paradigm shift renders it irrelevant or, I suppose, that the world comes to an end.

For an example, we again consider the meeting of the Bunsen Society in Hamburg in 1907. Assuming the role of a *tertius gaudens* broker, Walther Nernst attempted to convince the assembled scientists that the equilibrium of ammonia was such that industrial upscaling was unlikely—and this after he had already entered into a contractual agreement with Grießheim-Elektron the year before to commercially develop this same process. Nernst's actions, after consideration of the experimental and theoretical numbers at his disposal, make it likely that he suspected from the earliest stages that industrial upscaling was actually worth pursuing. However, from a scientific perspective, this conclusion is immaterial: either the rules of physical chemistry allow for the industrial synthesis of ammonia using current technological capabilities or they do not. At most, any "success" Nernst could have had as a *tertius gaudens* broker would have been to detract others so that he could benefit financially or receive recognition for being the first to complete the synthesis on a laboratory scale. It is certainly possible that he had such motivations; Nernst was a business man and did not shy away from vanity. Rudwick also discusses the career acumen and financial means of the actors in the Devonian Controversy and comes to the conclusion that they had no effect on the outcome (Rudwick 1985, pp. 438–445).

The instances of conduit, *tertius iungens*, and *tertius gaudens* brokerage employed during the scientific development and early-stage industrialization of ammonia synthesis are shown in Fig. 18.1. Both Haber and Nernst pursued a combined conduit-*iungens* strategy of brokerage when bridging the gap between

Brokerage Type	Fritz Haber	Walther Nernst	Carl Bosch	Alwin Mittasch
Conduit	Knowledge transfer from theory to lab (assistants) requiring known/feasible technology	Knowledge transfer from theory to lab (assistants) requiring known/feasible technology	Knowledge transfer from lab to industry requiring known/feasible technology	Knowledge transfer from lab to industry requiring known/feasible technology
	Knowledge transfer from lab to industry requiring known/feasible technology	Knowledge transfer from lab to industry requiring known/feasible technology‡		
tertius iungens	Knowledge transfer from theory to lab requiring new technology*	Knowledge transfer from theory to lab requiring new technology†	Knowledge transfer from lab scale to industrial scale requiring new technology	Knowledge transfer from lab scale to industrial scale requiring new technology
	Publication of experimental/theoretical results	Publication of experimental/theoretical results		Knowledge transfer between scientific paradigms Empirical ↔ Physical Chemistry
	Knowledge transfer from lab scale to industrial scale requiring new technology	Knowledge transfer from lab scale to industrial scale requiring new technology‡		
tertius gaudens		Supported perception that ammonia synthesis would not work on industrial scale while attempting to achieve it		

*Fritz Haber's technical assistants Gabriel van Oordt and Robert Le Rossignol played active roles

†Walther Nernst's technical assistants Karl Jellinek and Friedrich Jost played active roles

‡Nernst's attempt at industrial upscaling was unsuccessful

Fig. 18.1 Brokerage behavior, some resulting in epistemic knowledge transfers, in the scientific breakthrough leading to the Haber-Bosch process

theory and their technical assistants in the realm of experiment and later, between science and industry (Obstfeld et al. 2014). One good example of the evolution of their approach is illustrated by the vessel in which ammonia was synthesized. Between 1905 and 1908, the synthesis was performed in a pressure oven with a catalyst to establish a precise temperature and pressure at which small amounts of nitrogen, hydrogen, and ammonia were in equilibrium (Part II, Chap. 11). The results, obtained by Haber, Nernst, and their assistants, were only a verification of physicochemical principles established by linking theoretical and laboratory considerations. In 1909, Haber and his assistant Robert Le Rossignol translated this information into a more industrial language by building their final laboratory apparatus, which included all of the aspects needed for industrialization (continuous ammonia production, heat exchange) but still at a laboratory scale (Part II, Fig. 12.1). The translation continued as Carl Bosch and Alwin Mittasch, with considerable effort and setbacks, upscaled the machinery and catalyst according to the principles established by Haber (Part I, Chap. 7). Only after those challenges had been overcome, could industrialization be completed with the factory at Leuna in 1913. In this final phase, Haber was absent.

Another, purely scientific, brokerage dynamic is found in the interaction between Fritz Haber and Walther Nernst themselves between 1903 and 1908. Here, they also brokered between theory and experiment with a combined strategy, albeit in ways more explicitly dependent on their backgrounds. Nernst brokered mainly from theory to experiment, while Haber maintained a balance, moving at times in either direction. One notable aspect is that some of this brokerage involved a two-step bridging of the structural holes (Fig. 18.2a). Theoretical knowledge sometimes moved from Nernst to Haber and then from Haber to experiment, or laboratory experience moved in the opposite direction. From another perspective, Haber and Nernst simultaneously occupied positions in both theory and experiment. The historical record does not allow for a full reconstruction of the state of physical knowledge of each man, but their specific behavior indicates their interaction may also be characterized via the pathway of knowledge transfer called the *structural fold*, shown in Fig. 18.2b (Vedres and Stark 2010). In addition, Fig. 18.2 shows Friedrich Jost as an independent contributor to research on ammonia synthesis after 1907 (Part II, Chap. 11).

It is also worth mentioning that brokerage can be used in the development of new talent for the next generation (Sgourev 2015), a pursuit to which Haber and Nernst were committed throughout their careers.

The analysis of Haber and Nernst's brokerage roles gives theoretical underpinning to the two scientists' abilities to shift between experiment and the mathematically demanding theory of physical chemistry. In Part II, I assigned great importance to this movement as a determining factor in Haber's breakthrough. Their positions, in some ways peripheral with ties to outside groups, are described as "structurally contradictory," "anomalous," or "stylistically incoherent" (Padgett and Ansell 1993; Sgourev 2015). Haber was the more anomalous of the two. His theory work may not have been as bold as Nernst's, but he was more than a proficient theorist. It was Haber's experimental expertise—or perhaps his attention to experimental detail—

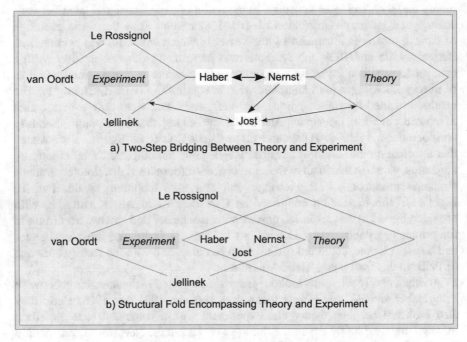

Fig. 18.2 Haber and Nernst's brokerage links between theory and experiment represented as (**a**) a two-step process bridging a structural hole and (**b**) a structural fold. Friedrich Jost's independent role is also evident

that gave him the decisive advantage over Nernst when he correctly incorporated the effect of the catalyst into his work. As he himself put it: "...a combination of experimental success with thermodynamic considerations was needed (Haber 1920, p. 326)." Haber's experience from his early career in the experimental domain gave him a novel perspective when he transitioned to theory, as did the ability to develop conceptual tools not available to the common scientist (Padgett and McLean 2006).

Nernst, through his relationship with Grießheim-Elektron, had a further intermediary, or structurally anomalous position, this time between science and industry. It is a position Haber would later assume with BASF and demonstrates the potential of bridging this kind of divide. Such brokerage activity may cause investments to be made in scientific research in a speculative way, or at the "low end" of the market, and may effect the speed with which a scientific discovery is industrialized, or who reaps the financial rewards. However, while these anomalous positions may be crucial in the transformation of scientific value into commercial market value, the resulting activity does not alter the scientific value itself.

The result of successful brokerage is not only knowledge transfer and coordination, but also the creation of a tie where a structural hole used to be (with respect to an epistemic transfer I used the term pathway). These ties can be strong or weak (or *perceived* to be such) and bind an alter, or actor, into a social network (Granovetter 1973; Ibarra and Andrews 1993; Perry-Smith and Shalley 2003). The strength of the ties, as well as to whom the alter is connected, help to indicate a central or peripheral position. Pioneers or pathfinders, whom we focus on here, often are (or perceive themselves to be) marginal. The classic conundrum is that innovative players at the periphery are free to act as they wish, but have limited influence. Influential players who are well-connected, on the other hand, are less free to act. Having weak ties is vital to being innovative, but there is a limit, after which an increasing number of weak ties becomes detrimental (Granovetter 1973; Kerckhoff and Back 1965), (Becker 1984, pp. 233–246), (Sgourev 2013, 2015). An alter's position is, furthermore, dynamic (Rudwick 1985, p. 420). Initially, "weak [boundary-spanning] ties are better than strong ties for creativity and. . .a peripheral position with many connections outside of the network is likely to be associated with more creative insights and potentially groundbreaking advancements" because "exposure to a new process of working or a new approach to a problem may serve as a seed that causes one to pursue previously unexplored directions. . ." However, success and exposure will lead to a new equilibrium and "eventually, the person will become so central in the network that he or she will become too entrenched or immersed, ultimately constraining creativity (Perry-Smith and Shalley 2003)."[2] Returning to the social network of a scientist, a setting in which access to knowledge of current trends is not necessary for success and where disregard for external public or professional opinion may have no consequence, we can see how the problem may be overcome of remaining innovative while occupying a central position. A scientist's social setting consists of colleagues and paradigm (the body of theoretical and experimental knowledge), and being "central" or "peripheral" can be with respect to either. A senior scientist can be well-connected (in, say, a political sense) to colleagues in the field of specialization but remain an isolate with respect to the paradigm; the ties to dead colleagues remains weak. This researcher is free to innovate and has the influence to publicize any results. Such a position is often assured via the power of job security (tenure) and research freedom, equating to the ability to take risks. It is also possible that a scientist, senior or junior, has many weak ties to living colleagues but chooses to stick closely to the chosen paradigm. It is unlikely this researcher will be innovative. In a unique situation, a researcher may be peripheral with respect to both living and dead colleagues. If the novel, creative work of this researcher proves correct according to the "objective" truth within the paradigm, the results will eventually gain acceptance.[3] While weak ties to living colleagues can be beneficial in science for identifying and articulating knowledge, they may also be limited; weak ties are not necessarily decisive for diffusion of knowledge through social contacts because the published scientific literature (especially renowned journals) can replace them. However, the number

---

[2] Fleck argued that this move from the periphery toward the center is accompanied by a of loss of *ability* to think outside of the paradigm (Fleck 1980, pp. 109, 121–124). While this assertion may be true in some cases, it is not a general characteristic of becoming more incorporated into a scientific community. However, the situations in which one can comfortably express one's self freely may become rare.

[3] The question of whether this researcher will receive due credit is another matter.

of ties can effect the speed at which the information disseminates. Strong ties to living colleagues are the key to amassing political clout, which may be needed for subjective decisions, such as setting research agendas or establishing scientific consensus (and are especially important if the choices are poor).

Part II, Chap. 10 describes how this social structure played into Haber's rationale for taking up the challenge of ammonia synthesis. Politically, he was a central player with job security, at least after he became a professor in Karlsruhe. However, as he had one foot in both theory and experiment, he had weak ties to both these sets of knowledge compared to many of his colleagues who had one main area of expertise. He was in a position to be innovative and to take a risk. As for Nernst, he had stronger ties to theory than Haber, but weaker ties to experiment. While also an advantageous position, in this case, Haber had a more beneficial mixture.

The discussion thus far has covered the transfer of knowledge within a paradigm, that is, during times of normal science. However, the example of ammonia synthesis shows transfers of knowledge between paradigms can also be decisive for a scientific breakthrough. Let us recall the aforementioned example of Haber facilitating the knowledge transfer between the paradigm of empirical chemistry (the qualitative behavior of the catalyst) and the quantitative mathematical theory of physical chemistry, at times with the help of Nernst. Another example can be found during the early stages of industrialization at BASF: the transfer of knowledge between Alwin Mittasch, who led purely empirical investigations to identify a suitable industrial catalyst, and Carl Bosch who was upscaling Haber's laboratory apparatus while remaining within the conditions dictated by physical chemistry. This brokerage dynamic is actually more complicated in that Bosch formed the link directly to engineering while Haber remained involved in the upscaling as the direct link to physical chemistry. We still see this activity today in scientific research. Investigations of catalyst materials, for example, involve multiple independent fields such as empirical chemistry and materials science, optics, the wave theory of light, semiconductor theory, thermodynamics, ab initio density functional theoretical calculations (quantum mechanics), classical electrodynamics, and relativistic electrodynamics (Greiner et al. 2018). The exploitation of these weak ties is reflected in publications with long author lists representing many institutes. The discussion of cross-paradigmatic work is concluded in Chap. 21.

One final, seemingly simple factor central to successful brokerage is that capable individuals need a reason to engage in knowledge transfer. They need motivation to interact, access to one another, and a working atmosphere that encourages and facilitates trust and collaboration (i.e. a research facility) (Brooks 1994; Pavitt 1990), (Csikszentmihályi 1996, chapter 6), (Burt 2004; Fligstein 2001; Obstfeld et al. 2014; Sgourev 2015). These factors seem so obviously important but are often neglected in practice. Silicon Valley, for example, has put much effort into cultivating and maintaining such an atmosphere (Part I, Chap. 8). In science especially, where value can be difficult to assess, sources of motivation can be ephemeral. Whether it be the pursuit of an aesthetic theory, a professorship, "the need to know," or the thrill of discovery, there are many genuine sources of motivation in science. These must be strengthened and proliferated because the structure of academic research can lend itself to attempts at *gaudens* brokerage—a choice that at best slows, but can also damage the scientific endeavor.

Chapter 19
Between Science and Industry: The Stage of 10–100

Here we consider one type of the dissemination of knowledge originating from science. Beginning with basic scientific research we can look in two directions. One is more fundamental, mathematics. The scientific perspective is one in which the realities of the physical world are used to consider the solutions offered by purely mathematical expressions (equations). If we look in the applied direction, we come to technology and industry. The transition to this stage, assuming the scientific discovery leads to a new technology, is difficult to investigate because it contains a mixture of the elements we have discussed thus far: the dynamics of science and the dynamics of technological innovation and commercial production. A description of this evolution using social network theory has, until now, eluded researchers because, it is reasonable to assume, the difficulties in understanding the wide array of necessary concepts are still too high from both sides. Network and innovation researchers are often not well-versed in scientific details and natural scientists rarely have experience outside of their immediate research field, much less in reflecting on the nature of the activities of their own profession. The starting point in the following is again the natural sciences, although the technological perspective is also valid. Such a mixture does not have to be viewed only as a transition. The discussion is applicable to the static assemblage represented by "useful knowledge" or "useful science" (as opposed to "scattered bits of knowledge") from the eighteenth century, the technological sciences, and ultimately the technosciences (Klein 2016b,c, 2020; Landecker 2008).

Although the exact nature of the transition has not been identified, we know it is there. We can easily find knowledge from the realm of pure science that has made its way into a technological-industrial setting. Can we use what we have covered thus far to conclude anything about the adaptor between the two?

One benefit of the language of technological innovation to investigate the scientific endeavor is that it allows us to consider the transition between the two. Instead of only a comparison of science and technology, we can analyze the point at which knowledge moves from the hands of scientists at the laboratory level to the

© The Author(s) 2022
B. Johnson, *Making Ammonia*, https://doi.org/10.1007/978-3-030-85532-1_19

Fig. 19.1 A double-cone structure modified to include the stage of 10–100

hands of industry with the goal of upscaling. It is not a sudden transition, rather the two branches overlap in the stage of 10–100.

Referring to the schematic diagram of the Haze in Fig. 17.2, we can resolve the diffusion of resulting technologies (the expanding cone) into two sections: a transitional and a main industrial stage (Fig. 19.1). The breakthrough stage linking the two cones contains perhaps one to 10 core individuals (scientists), whereas the successful realization and management of full industrial upscaling consists of perhaps 100 to 1000.[1] We see these two stages clearly when considering ammonia synthesis. The main players involved in the scientific development are found in Fig. 18.2. There are 6 in total (7 with Wilhelm Ostwald), whereas in 1927 and 1928 after industrial production 35,000 people worked at the ammonia synthesis plants at Oppau and Leuna (Abelshauser et al. 2004, p. 224). Between these two phases is the transitional stage, containing perhaps 10 to 100 individuals, in which scientists can have a central role in initial upscaling efforts (Obstfeld 2019).[2] With ammonia synthesis, this stage began at BASF in 1909 when Haber and Le Rossignol demonstrated their working prototype and ended in 1911 after Bosch had solved the problem of the degradation of steel pipes (Bosch holes) and Mittasch had developed an economical and effective catalyst (Part I, Chap. 7 and Part II, Chap. 12). Here, both research scientists and engineers were involved in moving the synthesis of ammonia (originally only on the laboratory scale) to the industrial scale. Not only were more individuals involved than during the breakthrough and far less than during full industrialization, but one may begin to speak of technological innovation

---

[1] This is an attempt at generalization. Today's "Big Science" shows that under some circumstances this breakdown may be more complicated. Also, small startups or information-based companies can operate at full capacity with small numbers of workers.

[2] Prof. David Obstfeld at the California State University, Fullerton drew my attention to these orders of magnitude.

despite the residual scientific influence. The work of both Bosch and Mittasch was guided by the conditions defined through the theory of physical chemistry and the results of Haber, Le Rossignol, van Oordt, Nernst, Jost, and Jellinek. Of these researchers, the first two were directly involved in the transitional stage at BASF, although only briefly.

Figure 19.2 shows the progression of combinations of knowledge surrounding ammonia synthesis up until the point it entered the transitional stage of 10–100. The first investigations on ammonia were empirical in nature with little theory to support the results. An initial framework was offered in the middle decades of the nineteenth century as early structural chemical theories and concepts of catalysis emerged, but the state of knowledge was not sufficient to understand how nitrogen and hydrogen united to form ammonia (Part I, Chaps. 3–6). Nevertheless, some empirical work began to resemble more modern experiments. It was only after about 1880 that the theory of physical chemistry was mature enough to offer real insight into the ammonia synthesis reaction. From our perspective today, we can see that attempts to synthesize ammonia at this time were bound to fail; there was no possibility of successfully bridging the gap between theory and experiment. The adequate combination of knowledge was first presented by Wilhelm Ostwald in 1900 when he wrote the "recipe" for the reaction by including all necessary theoretical and experimental factors. While his attempts also failed (though not because of any glaring conceptual oversight), they set the stage for the efforts of Fritz Haber and Walther Nernst. The interaction between these two scientists brought significant consistency between theory and experiment as they brokered the knowledge transfer that led to a successful laboratory and ultimately economically viable industrial process. The upscaling was carried out between 1909 and 1911 at BASF, where Haber was joined by Carl Bosch, Alwin Mittasch, and their assistants, forming the stage of 10–100. The development represented in Fig. 19.2 is a consequence of successive actors occupying more structurally anomalous roles. Their positions can be described increasingly as peripheral and finally as occupying bridging positions or structural folds. The transition was due to each individual's collection of experience and knowledge and also to a reshuffling of groups (Vedres and Stark 2010). One example of the changing roles is found in the patent disputes in 1910 where Walther Nernst worked with Haber on the side of BASF, functioning as an expert witness instead of a broker between theory and experiment (Part II, Chap. 12). Another example is illustrated by the early attempts to synthesize fixed nitrogen at BASF at the turn of the century. At that time, ammonia synthesis from the elements was considered a fringe possibility and Carl Bosch and Alwin Mittasch had concentrated on the multi-step cyanamide process and the electric arc where they gained experience in the properties of mixed materials, the thermodynamics of chemical reactions, and the required technical equipment. Fritz Haber was involved in some of these investigations, but other actors were more integral. After 1909, the core of the fixed nitrogen group—now focused on Haber's method of direct ammonia synthesis—was rearranged to contain not only Haber (Le Rossignol left for a job in Berlin in August of 1909 (Sheppard 2017)) but also Bosch and Mittasch. By that time, the three had already amassed considerable knowledge from different

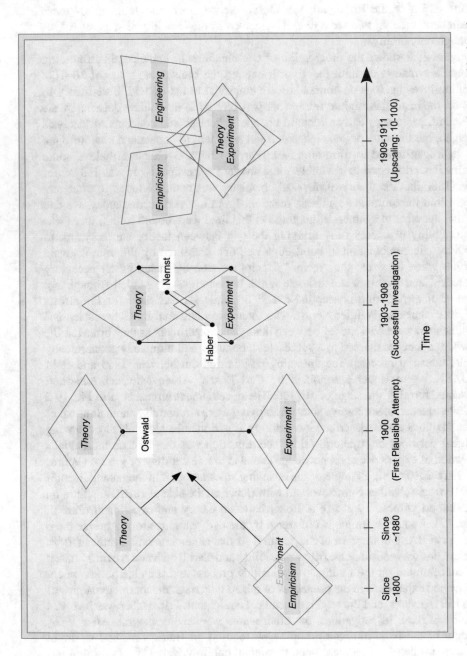

Fig. 19.2 The relationship between experiment and theory on ammonia synthesis. Scientific progress results in improved consistency and leads to the stage of 10–100. New experience and cooperation results in a tendency toward more structural fold-type relationships. Political and engineering (subjective) decisions as well as *tertius gaudens* brokerage begin to effect the final outcome of decisions

disciplines. This outcome is represented in Fig. 19.2 by the right-hand configuration where the question of theory and experiment on the scientific level was settled and engineering questions (Bosch and an industrial-scale synthesis facility) and empirical investigations (Mittasch and the catalyst) are placed so as to remain within the boundary conditions dictated by physical chemistry.

What is apparent in the stage of 10–100 is that political and engineering decisions as well as *tertius gaudens* brokerage no longer only effect the speed of progress as they did in the purely scientific stages. Out of an array of possibilities, of which any single one may appear subjectively "superior" to the others, it is possible to make particular choices as to the materials used or the layout of the facility; all these decisions effect efficiency and ease of operation. The objectively "correct" choice may neither be identifiable nor even need to be found. Many different configurations are adequate. The result is that aesthetics, opinions, or desires now have the potential to influence the decision making process with real and lasting consequences for the final outcome. The stage of 10–100 is where we stop waiting for answers to be revealed and start actively designing what we need.

The upscaling of ammonia synthesis at BASF has the potential to further clarify the stage of 10–100. However, for this research, Fritz Haber's personal network based on a survey of archival materials, particularly his private correspondence and his communication with BASF, is required.

Chapter 20
Risk and the Acceptance of Failure

In the end of Part I, I discussed another aspect of advancement: the willingness to take a risk in order to increase the likelihood of making considerable progress but also of coming up empty handed. Through the Silicon Valley Model, for example, industrial players have been able to incorporate risk taking and the acceptance of failure into their business models. It is not only in the tech industry that we see the repercussions. In the art world, too, risk is a sticky thing and requires systemic support to bolster change (Becker 1984, chapter 4). Considering the emergence of Cubism in Paris around 1910, for instance: "As the costs of experimentation were suddenly reduced and dealers began to assume the risk of failure, the preconditions were created for the pursuit of art that was not simply different, but radically so (Sgourev 2013)."

In fact, there is a broad consensus in the literature that risk plays a key role in change and innovation processes across many disciplines (Rosenberg 1983, p. 289), (Csikszentmihályi 1996, pp. 257–258), (Bonvillian 2014; BUND 2012; Burt 2004; Graßhoff 2008; Obstfeld et al. 2014; Sgourev 2013, 2015).

Science, however, has remained more conservative. One reason is the enigmatic nature of scientific value; the arrival of a new idea or product does not have the same power to drive science as it does innovation in the market place. Therefore, pressure to perform normal science comes from personal initiative or from the structure of the system. Only those with a secure yet peripheral position (with respect to the paradigm) can be creative and afford to take risks. In contrast to industry, the number of these individuals is limited in scientific research; young scientists (and academics in general) are actually advised to limit risk wherever possible. Yet science, too, despite its propensity for conservatism, has been shaped by the positive and negative consequences of risk taking. As Max Planck stated in 1913 in reference to the work of Albert Einstein, "it is...not possible to implement novelty in the exact natural sciences without daring to take a single risk (Planck

© The Author(s) 2022
B. Johnson, *Making Ammonia*, https://doi.org/10.1007/978-3-030-85532-1_20

et al. 1913)."[1] Haber took a risk when he began research on ammonia synthesis in 1903 and also when he continued to work toward industrial upscaling in the face of contemporaneous opposition. It seems we would be well advised to cultivate (and possibly increase) the acceptance of high-risk endeavors in science because the Haze is not an enduring entity. Vannevar Bush noticed this aspect in 1945: "...there is a perverse law governing research: under the pressure for immediate results, and unless deliberate policies are set up to guard against this, *applied research invariably drives out pure* (Bush 1945b, pp. *xxvi*)." And what is basic research if not the acceptance of the risk of finding something of little or no scientific value?

A new structure of our research environments—here I broaden the term research to include not just the natural sciences but all of academic research—is needed if risk taking is to be accepted and promoted. The problem is that rewarding risk in research is not an easy task, for what is the real value of an undertaking that has "failed"?

Today, there is a particular problem solving approach that has been awarded much attention without receiving corresponding opportunity (Bromham et al. 2016). It is the interdisciplinary approach (with transdisciplinarity afforded no better standing) (Scholz 2001, chapter 15). The lack of acceptance of projects filed under this rubric has been attributed to, among other reasons, the inability of a single review panel to properly assess its significance or that such proposals are too high risk. Currently, there certainly are risks of failure associated with an interdisciplinary research approach. However, I argue this concern is more a consequence of lack of proper strategy and understanding of how interdisciplinary collaboration differs from "traditional" research and less an inherent weakness of interdisciplinary work itself. For perspective, we can tie together several aspects of the Haze. In the introduction, I touched on the use of terminology and how it is important to use language and concepts in a way that remains true to their definition. Communication is straightforward if there is a common lexicon, which, almost by definition, is not the case in interdisciplinary work. What is required, then, to surmount this boundary? An initial strategy seems to require nothing more than the obvious, yet often neglected, factors stated at the end of Chap. 18: motivation to work together, access to one another, and a working environment that promotes collaboration. To these, one more attribute must be added: patience. It takes time to develop a shared lexicon, but it is entirely possible. If communication difficulties arise between experienced professionals, it is not because one side is too dumb, but rather that they do not understand each other. Most everyone is a layman outside their own area of expertise, making collaboration a two-way, yet uneven street. Everyone exposed to an interdisciplinary working environment will eventually find themselves in the position of the outsider. Or of the failure. This is the risk, and it should be viewed as something of vast potential instead of an irreparable defect.

Case in point: this book would not exist if not for a shared mindset toward risk taking and patient, supportive interdisciplinary collaboration from my colleagues.

---

[1] "...ohne einmal ein Risiko zu wagen, läßt sich...in der exaktesten Naturwissenschaft keine wirkliche Neuerung einführen."

Chapter 21
The Haze: Interdependencies in Science

The discussion of the Haze so far has only peripherally included the bifurcation of science into "normal" and "crisis." This perspective has been useful in illustrating some aspects of scientific research. However, the complexity of the Haze along with the nature of paradigms may have already made it apparent that the notion of a parallel existence of two distinct kinds of science is too simplified, especially when considering modern mathematically based theories and the advanced technology needed for the production of supporting experimental evidence. The success of the scientific endeavor depends on an interdependence of "normal science" and multiple paradigms so that there is no clear separation between fact-gathering and "times of crisis" in science leading to paradigm change (whatever form the latter may take). Both are always present in some form and to some degree. The consideration of an example of modern research provides a succinct illustration.

Here again, a concept is borrowed from innovation research. Dominant design is a theory in which an overarching plan governs the central idea of different technologies arranged in a hierarchy of nested subsystems (Abernathy and Utterback 1978; Murmann and Frenken 2006; Sanderson and Uzumeri 1995).[1] Instead of a technological assemblage, I illustrate a successful scientific experiment and its reliance on "normal science" as well as different paradigms. The paradigms are linked directly to the "core components," or fact-gathering activities, within the hierarchy of basic building blocks of today's investigations of complex systems. Figure 21.1 shows the example of the optimization of catalytic materials with a synchrotron-based X-ray source (Greiner et al. 2018). The mixture of methods and corresponding paradigms needed to interpret experimental results and characterize system behavior—empirical chemistry and materials science, solid state theory, optics, the wave theory of light, statistical mechanics, thermodynamics, classical

---

[1] Dominant design has been likened to Kuhn's notion of a paradigm (Murmann and Frenken 2006). While it is a promising approach—relationships between paradigms identified via certain similarities—here the aspect of dominant design is emphasized that illustrates the interconnection of parts of a complex system.

© The Author(s) 2022
B. Johnson, *Making Ammonia*, https://doi.org/10.1007/978-3-030-85532-1_21

electrodynamics, quantum mechanics, and relativistic electrodynamics, not to mention engineering abilities—are depicted as a four-level nested hierarchy. The paradigms are embedded in a way specific to each subsystem that ultimately defines the systems level. Their distinct arrangement is advantageous for certain studies so that research goals and individual components (theory or measurement methods and, therefore, paradigms) may be exchanged for adaption to different experiments. For example, slight changes in components 1–4 would lead to an effective regime for optimizing materials for photovoltaics. Like dominant design technologies, the higher the sub-system level that is replaced, the more effort is required to make the changes.

Figure 21.1 explicitly reflects the complexity of the Haze, now comprising multiple paradigms, and a way in which its condensation may be illustrated. This outcome is natural. The scientific paradigms (or theories) currently available to us are each limited in the behavior they can describe while real physical systems are not subject to the same constraint. In order to understand these systems, it is not sufficient to view the universe in only one way. A set of different paradigms is required to solve scientific problems, whether "normal" or of the "crisis" variety. Furthermore, the creativity inherent in science along with a unique aesthetic value becomes apparent in the hierarchy. It shows the strict, overriding rules often associated with scientific research to be infused with a subjective element of strategy that is needed to overcome the complexity. Until now, the words "innovation" or "innovative" have been avoided with regards to science; instead the words "discovery" and "breakthrough" have been consistently used.[2] Yet Fig. 21.1 makes it difficult to completely exclude a sense of innovation, if only with respect to the artistic traits of scientific research.

This interplay begs the question of the communication (or brokerage) between paradigms themselves. "Communication across the revolutionary divide is inevitably partial," wrote Kuhn. Later he continued, "before they [scientists] can hope to communicate fully, one group or the other must experience the conversion that we have been calling a paradigm switch (Kuhn 1970, pp. 149–150)." Similarly, Fleck wrote, "the greater the difference between two thought styles, the more inhibited will be the communication of ideas (Fleck 1979, p. 109)."[3]

Much has been made of the veracity of these statements, but it is apparent that the ability to communicate is *adequate* for successful cross-paradigmatic work. This is as true today as it was in Haber's time and as it was in the eighteenth century. Antoine Lavoisier, while conducting his combustion experiments on sugar in 1787, used the "pre-chemical revolution"[4] nomenclature, "vitriolic" and "inflammable air," and "post-revolution" terminology, "sulfuric" and "hydrogen gas." "If it is true," wrote Lawrence Holmes, "that revolutionary changes in science imply new uses

---

[2] Rudwick prefers the term "shaping" of knowledge (Rudwick 1985, p. 15).

[3] "Je größer die Differenz zweier Denkstile, um so geringer der Gedankenverkehr (Fleck 1980, p. 142)."

[4] See footnote 2 in Part I.

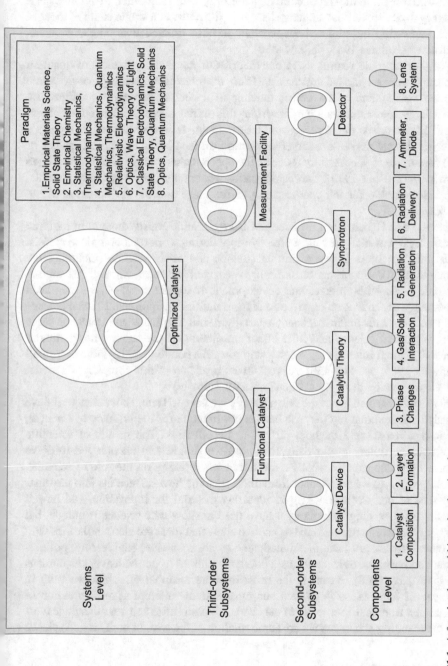

Fig. 21.1 A four-level nested hierarchy showing the components and subsystems required to achieve an optimized catalyst using synchrotron-based X-ray emission. Small changes in 1–4 would allow for the optimization of materials for photovoltaics. Inset: the paradigms critical to each component. Based on fig. 1 in Murmann and Frenken (2006)

of language which make it impossible[5] for those who have adopted the new and those who remain in the old to communicate fully with each other, then Lavoisier's tendency to cross over that language barrier in the midst of a single chain of thought raises interesting questions about whether he also crossed over in the thoughts themselves (Holmes 1985, pp. 334–336)."

Another example returns us to our theories of mechanics. A modern quantum mechanist and a classical mechanist (from the nineteenth century) could agree that the stable state of an object moving in a central potential is defined by total energy and rotational inertia. After this initial accord, a heated argument would likely break out, but it would be enough to move beyond a layman's conversation and maybe even to collaboration. In fact, Niels Bohr used aspects of both paradigms to successfully describe the spectral characteristics of the hydrogen atom in 1913 (Bohr 1913). Interparadigmatic communication is not only possible, it is indispensable for the condensation of the Haze and the precipitation of the breakthrough.

I argue that Haber, like Lavoisier and Bohr, simultaneously thought in multiple languages; it was likely his most effective tool during work on ammonia synthesis. Thus, the line blurs between normal science and crisis science (including the paradigm shift). Important breakthroughs as well as large amounts of routine scientific work—the progress and development of science itself—have long been consistently dependent on contributions from multiple paradigms. Normal science and paradigm shifts are interrelated; a paradigm shift may have vast effects on some areas of normal science and little effect on others. At the same time, particular results of normal science may mean a step toward a paradigm shift, large or a small (Rudwick 1985, pp. 448–450). A breakthrough, as part of this complex of activity, is made distinct by the context of an arena for discovery.

Scientific research is not the only strategy we have to bring about change (I have already indicated that my purpose here is not to argue for the primacy of science), but it is a remarkable method, proven to be effective. The results of scientific investigations influence our daily lives and we should feel encouraged to delve more deeply into exactly what we are doing when taking on the role of scientist. Not only that, but we will benefit from understanding how science fits into the other processes of our world, how it provides raw material for innovation, and how it benefits from existing technology. I have tried to show why science is unique, but another objective in this section has been to show that there are also similarities with the other endeavors I have mentioned such as art, medicine, engineering, politics, sports and business. Science forms a bookend with art to our endeavors as humans (de Santillana 1968). As art is the most extreme example of subjective truth in the eyes of humans, so is science our most extreme attempt at an expression of objectivity. Both make sense out of the chaos of possibilities and our other endeavors and the resulting achievements are some mixture of the two. In both art and science

---

[5] I find the word "impossible" too strong. Kuhn also defended himself against this binary choice (Kuhn 1970, pp. 198–199).

there is infinite choice. In art it is a continuous variation of anything we can render; in the natural sciences, there are discontinuous sets of rules to describe our universe according to how we are able, or how we are forced to view it. We may never find the end of either.

Chapter 22
One Final Element

There is one component that, but for a few intimations, has been conspicuously missing from the description of the Haze: the role of curiosity in scientific research.

Why was it not included as an element of fundamental significance? Is science not driven by deeply curious individuals striving for the best version of the truth they can describe, themselves compelled largely by passion or interest?

There is certainly a perception that curiosity is at the core of why one would choose, over the course of years or decades, to commit one's self to the pursuit of a seemingly esoteric tidbit of information. There are, therefore, several reasons to dispel this notion. At the end of the day, Scientists are people who must pay their bills, prepare dinner, and navigate life's snags just like everyone else. Society invests vast resources at all levels of research from basic to applied, and it is done with the expectation of a professional conversion of these means into results. Sometimes there is precious little room for curiosity because the problems are too complex and require mature, nuanced approaches to solve them. Scientists are, after all, not children. I do not mean this in a flippant way, as children have a reason to be curious: they are at a stage in life where their development is driven by determining what works and what does not in order to amass a body of tacit knowledge that allows them to function in everyday life. But children are not charged with making concrete contributions to society, and so they are just fine in that short, whimsical stint in life in which curiosity is at the center of so many things.

However, curiosity as the main driver of scientific activity is too idealistic.

Curiosity no doubt plays a role in science and is closely connected to creativity, especially in the *non-mission oriented research* of Chap. 17. This was illustrated concretely in Part II, Chap. 10, where Haber saw ammonia synthesis as having technological and market potential, but also as basic research contributing to the understanding of the free energy. It was something that fascinated him. There have been scientists—or natural philosophers—who could attribute a larger share of their

B. Johnson, *Making Ammonia*, https://doi.org/10.1007/978-3-030-85532-1_22

motivation to curiosity than many can today.[1] These were the few who had access to adequate and uncritical financial support and who had the correct disposition to be compelled by interest alone. Out of all of them, even fewer were able to employ this strategy throughout their careers. One may refer to them as lucky, not in the sense that they were allowed to be curious, but rather that several factors combined in a way that allowed them to practice science in the way best suited for them. Often though, the complexities of real scientific pursuits force compromise, no matter the researcher's disposition. Even today's top researchers do not have the luxury of exclusively following their own curiosity. There are so many factors leading to success that curiosity, in the end, should be considered a good (or even essential) motivator, or as a part of strategy to derive energy for the day-to-day challenge of staying focused. Curiosity can draw someone back to a subject, a sample, or a theory, but it does not represent a complete strategy for obtaining a full description of a system and formulating practical answers to basic questions about that system's behavior. There are too many critical details and high expectations which require less fascination and more drudge work, but which must be considered nonetheless. The merely curious person would simply neglect these tasks.

Science is, instead, driven by other factors: pride, a human need to know, the enjoyment and gratification of successful puzzle-solving, or the creation of elegance and beauty. There is also competition, which manifests itself internally (as a desire for renown or recognition as The One who made the discovery, and also through peer review and publication) and externally (through competition for limited funds and positions). Opportunities are kept at a critical level in research to afford a certain degree of knowledge generation while also narrowing the playing field to force out those actors who are not considered to be producing enough knowledge, or the right kind.

These can be summed up as ambition. Ambition to solve a problem or to gain insight. Or in some cases, to proliferate one's self and move up the hierarchy.

We often confuse this with curiosity.

To disentangle them we return to a central theme of this book: knowledge and how it is produced, moved, combined, and recombined. In science, an often meandering enterprise, curiosity alone will not produce a complete knowledge set. But ambition can if it is based on a sense of duty to solve a problem or gain insight. The word "ambition" often has a negative connotation, as if any success achieved need not be based on merit; a more positive word is "aspiration." However, ambition can be positive and we need ambitious research agendas and ambitious ideas to tackle those agendas. We just should not go too far because comparatively little knowledge will be produced if the only ambition is to receive recognition, especially in cases where the results only appear to be useful at first. Before we have the correct answer, things in science may seem to fit together in any number of ways,

---

[1] Today, the degree to which one may have the privilege to be driven by curiosity can be correlated to one's gender or family background (Middendorff et al. 2017; von Wensierski et al. 2015; Weininger and Lareau 2018).

due perhaps to the complexity of the problem, to limited contextual knowledge, or because someone has successfully convinced us that their answer is the correct one. To be sure, the latter has no impact on the inherent scientific value of that element of knowledge. As long as we continue the scientific endeavor, however, the pieces will eventually fall into place.

We may not be able to fully control the Haze, but we can tip the probability of condensation in our favor. Despite the random occurrence of decisive events in science, their outcomes can be understood in retrospect. Regarding past scientific discoveries, a nucleating effect can be derived from the knowledge base, especially in the hands of inventive or resolute minds, positioned and willing to take a risk. An active center is formed from the successful synthesis of ideas on which later concepts and conclusions can condense, accompanied and facilitated by the rigor of deduction and mathematical abstraction. We understand and can do these things well. It is only a matter of positioning ourselves at a point likely to seed the next discovery.

Appendix A
Approximations of Free Energy Fitting Functions

While examining Haber and Nernst's work on ammonia synthesis, it is important to consider how they interpreted and displayed their results. During experimental investigations, the two scientists and their assistants fit their data with approximations derived from fundamental physicochemical principles, which emphasize an important problem in physical chemistry: the determination of the change in free energy of chemical reactions (Bodländer 1902; Haber and van Oordt 1905b; Nernst 1903; van't Hoff 1898; von Jüptner 1904a,b,c,d). The equations presented here are related to the van't Hoff equation and describe the dependence of free energy changes on the equilibrium constant. They are derived without entropy, although alternative methods did employ it (Haber 1905b, pp. 27–54). This discussion is meant to do two things. First, it clarifies some of the physical assumptions contained in the fits (for example, in Figs. 11.7–11.11) and helps illustrate the scientists' approach to their data. In the original publications, the constants and units in the approximations are often not explicit so that they appear as arbitrary fitting functions. Second, this appendix and Appendix B (which addresses the derivation of the free energy and Nernst's heat theorem) present the mathematical context of thermodynamics and physical chemistry in which ammonia synthesis was studied at the time. Through this context, the critical role of the ammonia equilibrium measurements in the verification of theory is made more accessible (see also Chap. 13). It is not meant to be an exhaustive presentation of Haber's and Nernst's theoretical work; while my approach is from a modern perspective, my aim is not to address all points of criticism to their derivations. Rather, the goal is to present further aspects of their thinking to improve physical and historical insight.

The modern analogues to the equations derived here are differential equations of state for systems in which the number of molecules of each species may change via chemical reaction. This degree of freedom is accomplished by a corresponding shift in the chemical potential of each component, leading to changes in the total free

© The Author(s) 2022
B. Johnson, *Making Ammonia*, https://doi.org/10.1007/978-3-030-85532-1

energy available in the system. The modern expression for the internal energy, U, or enthalpy, H, along with the corresponding Helmholtz free energy, A, or Gibbs energy, G, can be derived starting with the first law of thermodynamics, which expresses the conservation of energy:

$$dU = d\,W + d\,Q \tag{A.1}$$

where dU is the change in internal energy, $d\,W$ the work done on or by the system, and $d\,Q$ the heat exchange. With the definitions of work performed during the compression/expansion of a gas in a hydrostatic system

$$d\,W = -P\,dV \tag{A.2}$$

and of entropy

$$dS = \frac{dQ}{T} \tag{A.3}$$

we arrive at

$$dU = -P\,dV + T\,dS \tag{A.4}$$

The definition of the Helmholtz energy is

$$A \equiv U - TS \tag{A.5}$$

so that

$$dA = dU - T\,ds - S\,dT = -p\,dV - S\,dT \tag{A.6}$$

Equations A.4 and A.6 describe a closed system in which the number of molecules of all chemical species remains constant. If, through a chemical reaction, the number of molecules of any constituent species changes, we allow for this shift by the addition of a term representing the reversible chemical work:

$$dU = -P\,dV + T\,dS + \sum_{j=1}^{N} \mu_j\,dn_j \tag{A.7}$$

$$dA = -p\,dV - S\,dT + \sum_{j=1}^{N} \mu_j\,dn_j \tag{A.8}$$

where μ_j is the chemical potential of the jth species and n_j the number of moles of that species (Zemansky and Dittman 1997, pp. 293–297, 400–403). Haber's and

Nernst's approximations have a similar term describing systems that may not be in equilibrium.

The derivation for enthalpy and the Gibbs energy are similar. The definition of the enthalpy, H, is

$$H \equiv U + PV \tag{A.9}$$

so that

$$dH = dU + P\,dV + V\,dP = V\,dP + T\,dS \tag{A.10}$$

The corresponding free energy, the Gibbs energy, G, is defined

$$G \equiv H - TS \tag{A.11}$$

so that

$$dG = dH - T\,dS - S\,dT = V\,dP - S\,dT \tag{A.12}$$

Allowing for the number of molecules of the constituent species to change through a chemical reaction:

$$dH = V\,dP + T\,dS + \sum_{j=1}^{N} \mu_j\,dn_j \tag{A.13}$$

$$dG = V\,dP - S\,dT + \sum_{j=1}^{N} \mu_j\,dn_j \tag{A.14}$$

Turning to the approximations for the free energy used in Part II, we begin with one derived by Haber in his book *Thermodynamics of Technical Gas Reactions* (Haber 1905b, pp. 1–26, 55–62) and also used in his first publication on ammonia in 1905 (Haber and van Oordt 1905b). Haber's original expressions are largely used throughout with some symbols and intermediate steps amended for clarity and continuity. Those experienced in modern thermodynamics may find particular aspects incomplete, however, in order to understand the scientists' arguments in Part II, it is important the original work be considered.[1]

Haber considered the differential expression for the free energy, A, describing an expanding ideal gas (Eq. A.15). This quantity represented the maximum work attainable from the system (free energy) because the internal energy of an incre-

---

[1] A modern and rigorous examination of the thermodynamic problem of ammonia synthesis can be found in (Müller and Müller 2009, chapter 9).

mentally expanding ideal gas remains constant. The temperature change caused by the heat flow, dQ, is exactly equal to the work of expansion, dW. The ideal gas law was then used to convert to molar volumes, v:[2]

$$dA = p\,dV = \frac{RT\,dv}{v} = RT\,d\ln v \tag{A.15}$$

or if integrated over an expansion from v to v':

$$A = \int_v^{v'} RT\,d\ln v = RT\,\ln\frac{v'}{v} \tag{A.16}$$

Haber reasoned that because this expression represented the maximum attainable work, it must be applicable to systems producing other forms of work, for example the electromotive force from a chemical reaction. If this were not the case a perpetual motion machine would be possible. "It remains only to show," he wrote (Haber 1905b, p. 20),

> that this relationship is not determined by the characteristics of the gases, but rather through heat and work, and remains valid for every latent heat as long as we limit ourselves to operations which deliver the maximum work during an isothermal process.[3]

He then expressed Eq. A.16 in terms of concentrations, c_j [mol/L], and employed a thought experiment known as van't Hoff's equilibrium box to evaluate chemical reactions in which the reactants and products are present in arbitrary (that is, possibly non-equilibrium) concentrations. Here, a compartment is envisioned in which the gases, say N_2, H_2, and NH_3, are always present in their equilibrium concentrations, c_j'. The reactants, nitrogen, and hydrogen, are introduced while the product, ammonia, is removed in such a way that equilibrium is maintained. If the reactants are present in equilibrium concentrations outside of the box, the process is not especially noteworthy. If, however, the reactants are present outside the box at some arbitrary non-equilibrium concentrations, c_j, they may be brought to the equilibrium concentration through an isothermal, path-independent process with the maximum work, A_j, before entering the box. Similarly, the product may be returned to the non-equilibrium concentrations through expansion or contraction (using an isothermal, path-independent process of maximum work) after removal from the box. The reaction within the box still proceeds at the equilibrium concentrations,

---

[2] Here, despite $(\frac{\partial A}{\partial V})_T = -p$, Haber left the value $p\,dV$ positive, although he referred to it as "the work performed (die geleistete...Arbeit)" by the gas. Furthermore, A is also a function of T, which Haber stated was to be held constant, but wrote dA as a full differential. Haber often used the full differential d instead of the partial differential ∂ when considering functions dependent on more than one variable. He only sometimes indicated which variable was to be held constant.

[3] "Es bleibt uns nur übrig, zu zeigen, daß diese Beziehung nicht durch die Eigenschaften der Gase, sondern durch die von Wärme und Arbeit bedingt ist und für jede latente Wärme, sofern wir uns auf Vorgänge beschränken, welche die maximale Arbeit bei isothermem Ablauf verrichten, Gültigkeit besitzt."

c'_j, but the overall result is the same as if the reaction had been carried out at the arbitrary concentrations, c_j, with the corresponding change in the maximum work obtainable from the system. Haber expressed this difference, A, with the sum[4]

$$A = A_1 + A_2 + A_3 + \cdots + A_N =$$

$$= RT \left\{ \mu_1 \ln \frac{c_1}{c'_1} + \mu_2 \ln \frac{c_2}{c'_2} + \mu_3 \ln \frac{c_3}{c'_3} + \cdots \mu_N \ln \frac{c_N}{c'_N} \right\} = RT \sum_{j=1}^{N} \mu_j \ln \frac{c_j}{c'_j}$$

$$(A.17)$$

where the μ_j are the stoichiometric coefficients, which Haber called "Molekülzahlen," and assigned them values depending on the number of molecules involved in the written reaction. The sign was determined by whether the species was a reactant $(-)$ or a product $(+)$. Separating the arbitrary and equilibrium concentrations into two terms, the expression became:

$$A = RT \ln \sum_{j=1}^{N} \mu_j \ln c'_j - RT \sum_{j=1}^{N} \mu_j \ln c_j = RT \ln K_c - RT \sum_{j=1}^{N} \mu_j \ln c_j$$

$$(A.18)$$

where K_c is the equilibrium constant expressed in terms of concentrations, or mole fractions and the first term represents the *standard free energy change* for a reaction taking place at equilibrium concentrations. The second term—which was to be measured—is analogous to the term added to the modern expressions above to describe systems whose numbers of molecules of any species can change via chemical reaction. Again, with help from the ideal gas law, Eq. A.18 was expressed as

$$A = RT \ln K_p - RT \sum_{j=1}^{N} \mu_j \ln p_j \qquad (A.19)$$

where K_p is the equilibrium constant expressed in terms of partial pressures, p'_j and p_j (see also Eq. A.28).

If the measured concentration is identical to the equilibrium concentration, then $\ln K_p = \sum_{j=1}^{N} \mu_j \ln p_j$ and $A = 0$. For Haber, **this was the equilibrium condition.** It was equivalent to stating (referring back to the van't Hoff box) that no additional work was required before or after the reaction because the constituents were present in their equilibrium concentrations.

---

[4] I have generalized Eq. A.17 for clarity—originally Haber illustrated the van't Hoff Box with the system $2H_2 + O_2 \longleftrightarrow 2H_2O$.

To develop an expression for the *standard free energy change*, Haber integrated van't Hoff's equation, in which the heat of reaction was for constant temperature and pressure. Multiplication with T yielded:[5]

$$R \int \frac{d \ln K_p}{dT} = - \int \frac{Q_{p,T}}{T^2} \quad \rightarrow \quad R T \ln K_p = C \cdot T - T \int \frac{Q_{p,T}}{T^2} dT \tag{A.20}$$

where C is the constant of integration.[6] Setting Eq. A.20 into Eq. A.19 resulted in

$$A = -T \int \frac{Q_{p,T}}{T^2} dT - R T \sum_{j=1}^{N} \mu_j \ln p_j + C \cdot T \tag{A.21}$$

$Q_{p,T}$ was expressed through the difference in average specific heats of the reactants and products between $0°K$ and T at constant pressure:

$$Q_{p,T} = Q_o + \rho'_p T + \rho''_p T^2 \tag{A.22}$$

Hans von Jüptner also used this expression and explained it in more detail (von Jüptner 1904a). Setting $Q_{p,T}$ into Eq. A.21 allowed for the integration of the expression:

$$A = Q_o - \rho'_p T \ln T - \rho''_p T^2 - \sum_{j=1}^{N} \mu_j \ln p_j + C' \cdot T \tag{A.23}$$

where C' is a new constant. This approximation was common for fitting and extrapolating measurements of equilibrium and was used in Haber and van Oordt's first paper on ammonia synthesis (eq. 9 in (Haber and van Oordt 1905b, p. 355) and Eq. (11.2) here in Sect. 11.1). It depended critically on the accuracy of the thermal data used to determine Q_o and the specific heats needed for the coefficients ρ'_p and ρ''_p. These established the value of the *standard free energy change*—the first term in Eq. A.19—and, thus, the precision of the equilibrium determination at $A = 0$. The term $\sum_{j=1}^{N} \mu_j \ln p_j$ was measured experimentally at a known temperature, T, and allowed the calculation of the constant of integration C'. Only then could the temperature dependence of the equilibrium mixture of the nitrogen-hydrogen-ammonia (or any other system) be plotted. It was one of the main experimental and theoretical difficulties in physical chemistry at the turn of the twentieth century, and

---

[5] This step has been simplified and makes use of the fundamental equation for the free energy $A - U = T(\frac{\partial A}{\partial T})_V$. See (Haber 1905b, pp. 60–61).

[6] There was no discussion of the variable-dependence of C.

the experimental challenges were on full display as thermal data and measurement accuracy were consistently at the core of Haber and Nernst's debate.

Another, simpler approximation was also used (despite Haber referring to it as "die gröbste Näherung"—the crudest approximation). Beginning with Eq. A.21:

$$A = -T \int \frac{Q_{p,T}}{T^2} dT - RT \sum_{j=1}^{N} \mu_j \ln p_j + C \cdot T \tag{A.24}$$

Bodländer, for example, simply chose (Bodländer 1902)

$$Q_{p,T} = Q = Const. \tag{A.25}$$

although it was known that the heat of reaction changed slowly with temperature and was described more accurately by Eq. A.22. However, the integrand in Eq. A.24 could now be easily evaluated:

$$A = -T \int \frac{Q}{T^2} dT - RT \sum_{j=1}^{N} \mu_j \ln p_j + C \cdot T = Q + \sum_{j=1}^{N} \mu_j \ln p_j + C' \cdot T \tag{A.26}$$

Setting $A = 0$ for equilibrium conditions and replacing the natural logarithm, $\ln$, with the base 10 logarithm, $\log$, the equation became

$$\sum_{j=1}^{N} \mu_j \log p_j = \frac{Q}{4.575 \, [cal/mol \, K] \, T} + \frac{C'}{4.575} \tag{A.27}$$

where again $\sum_{j=1}^{N} \mu_j \log p_j$ is the experimentally determined value, which at equilibrium is equivalent to $\log K_p$. The factor 4.575 results from multiplying the ideal gas constant, $R = 1.987$ cal/mol K, by 2.30, the result of changing logarithmic bases.[7] All terms in Eq. A.27 are dimensionless. As with Eq. A.23, knowledge of the composition of the equilibrium gas mixture at one temperature allowed for the determination of the constant, and an evaluation of the function. Equation A.27 was used extensively in both Haber's and Nernst's work.

A third, theoretical approach to determine the equilibrium between N_2, H_2, and NH_3 was derived by Nernst in 1906 (Nernst 1906) and is discussed with his results in Part II, Sect. 11.1.1 as well as in Appendix B.

These equations described the temperature dependence of the equilibrium position well near the temperature at which the measurement was made, but the

---

[7] As with the factor 4.575, when the calculations are carried out with a calculator today and compared to the numbers in the following sections, the influence of significant digits is evident.

approximations and uncertainties in thermal data—the heat of formation of ammonia and specific heats—caused deviations as the temperature decreased to a region
where catalysts were most effective and ammonia yields rose. For this reason, the
economical success of an upscaled ammonia synthesis remained dubious during the
initial studies. Modern calculations for ammonia equilibrium (filled black circles)
are given along with experimental results in Part II, Figs. 11.7–11.11, and illustrate
this shortcoming.[8]

At normal (atmospheric) pressures, only small amounts of ammonia were
obtained at equilibrium under the most advantageous conditions and even at higher
pressures, yields were not appreciably increased. To examine why, we consider an
equation which is helpful in understanding Haber and Nernst's results at pressures
above 1 atm. There are two forms of the equilibrium constant, one expressed in
terms of partial pressures, K_p, and one expressed in terms of concentrations, or
mole fractions, K_c. Their relationship in the ammonia-nitrogen-hydrogen system is:

$$K_p = \frac{p_{NH_3}}{p_{N_2}^{1/2}\, p_{H_2}^{3/2}} = \frac{c_{NH_3}}{c_{N_2}^{1/2}\, c_{H_2}^{3/2}}\, P^{1-\frac{1}{2}-\frac{3}{2}} = K_c\, P^{-1} \qquad (A.28)$$

where P is the total pressure and the relationship between the partial pressure and
the concentration is $p_x = c_x \cdot P$. Because K_p is independent of pressure, K_c
must compensate for changes in P by adjusting concentrations. According to Le
Chatelier's principle, higher pressures will cause the system (K_c) to adjust in a way
that favors the side of the chemical equation with the least number of moles. In
the case of $N_2 + 3\,H_2 \leftrightarrow 2\,NH_3$, the formation of ammonia is supported. We
see in Eq. A.28 that for small ammonia concentrations, as measured by Haber
and Nernst between 1904 and 1908, the denominator ($c_{N_2}^{1/2}\, c_{H_2}^{3/2}$) remains
essentially constant so that concentration changes are linear with the pressure, P.
However, once the ammonia concentration becomes appreciable, it will begin to rise
exponentially to offset the change in denominator (Haber and Rossignol 1908f; Jost
1908a), (Atkins 1987, pp. 226–231).

---

[8] The modern calculations were provided by Dr. Kevin Kähler at the Max Planck Institute for
Chemical Energy Conversion.

Appendix B
Theoretical Determinations of the Free Energy: Hermann von Helmholtz and Walther Nernst

In 1906, Walther Nernst published his article "Ueber die Berechnung chemischer Gleichgewichte aus thermischen Messungen" (On the calculation of chemical equilibria from thermal measurements) (Nernst 1906), which became the theoretical substantiation of Fritz Haber's experimental results. Here, as with Haber in Appendix A, we review Nernst's theoretical determinations of the free energy. As before, this presentation is not meant to fully clarify or criticize Nernst's approach to thermodynamics, but rather to reproduce his original calculations of chemical reactions in order to embed his thinking in the context of thermodynamics and physical chemistry. A contribution from Hermann von Helmholtz is also included, which Nernst used as the starting point for his own work.

The basic equation governing the relationship between the free energy, the total energy, and the entropy, $S = -(\frac{\partial A}{\partial T})_V$, of a system is, in differential form:

$$A = U + T \left(\frac{\partial A}{\partial T} \right)_V \qquad (B.1)$$

The integration of this function contains an unknown constant, $C(V)$,

$$A = -T \int \frac{U}{T^2} \, dT + C(V) \cdot T \qquad (B.2)$$

The approximations in Appendix A required an experimental measurement to determine the constant and calculate the absolute value of the function with temperature; the result was left burdened with experimental error. Walther Nernst's 1906 publication provided a solution to this problem, based on what would become known as the third law of thermodynamics (there was no mention of this name in Nernst's paper, as it was originally just a tool to help him solve a mathematical equation) and the thermal properties of the substances involved in the chemical reaction.

© The Author(s) 2022
B. Johnson, *Making Ammonia*, https://doi.org/10.1007/978-3-030-85532-1

As discussed in Part II, Chap. 13, the problem of the free energy grew out of attempts to predict the heat released during a chemical reaction. A quantitative explanation was published in 1878 by Josiah Willard Gibbs (1878a; 1878b) and in 1882 by Helmholtz (1882). Helmholtz used the first and second laws of thermodynamics to define a thermodynamic state function dependent on the state variables temperature, total (internal) energy, and entropy. This equation exposed an intimate relationship—or balance (Eq. (13.1))—between the energy and entropy of a chemical system. It illuminated not only a fundamental characteristic of nature but also confirmed the significance of entropy to a still unconvinced scientific community (Smith 1998, Chapters 8 and 11), (Müller 2007, pp. 148–149).

"We already know," wrote Helmholtz in 1882 (von Helmholtz 1882),

> that [changes in aggregate state or density] are capable of producing or consuming work, namely first in the form of heat, secondly in the form of other, unrestrictedly convertible work [...] it does not seem questionable to me that also in chemical processes the differentiation should be made between the portion of the power of chemical affinity that can be freely converted into other forms of work and the portion which can be produced only as heat. In the following, I will take the liberty to simply label these two kinds of energy the free and bound energy. Later, we will see that the processes which start from an idle state and proceed spontaneously at constant temperature without help from external work will only proceed in a direction that reduces the free energy...[It] is the values of the free energy, and not those of the total energy corresponding to the heat of reaction, which are decisive for the way in which the chemical affinity will act.[1]

Helmholtz identified two types of energy: the free energy (freie Energie), which could be fully converted into work and resulted from "ordered dynamics" within the system, and bound energy (gebundene Energie), which resulted from unordered dynamics. Here, a particle's motion was not required to have any relationship to that of its neighbor (Königsberg 1903). He derived mathematical expressions for his ideas by observing the total energy of a system as consisting of these two distinct parts.

---

[1] "[...wir wissen] schon, dass [Änderungen des Aggregatzustandes und der Dichtigkeit] Arbeit in zweierlei Form zu erzeugen oder zu verbrauchen fähig sind, nämlich erstens in der Form von Wärme, zweitens in Form anderer, unbeschränkt verwandelbarer Arbeit [...] so scheint es mir nicht fraglich, dass auch bei den chemischen Vorgängen die Scheidung zwischen dem freier Verwandlung in andere Abreitsformen fähigen Theile ihrer Verwandschaftskräfte und dem nur als Wärme erzeugbaren Theile vorgenommen werden muss. Ich werde mir erlauben diese beiden Theile der Energie im Folgenden kurzweg als die freie und die gebundene Energie zu bezeichnen. Wir werden später sehen, dass die aus dem Ruhezustand und bei constant gehaltener gleichmässiger Temperatur des Systems von selbst eintretenden und ohne Hilfe einer äusseren Arbeitskraft fortgehenden Processe nur in solcher Richtung vor sich gehen können, dass die freie Energie abnimmt...[es] würden also die Werthe der freien Energie, nicht die der durch Wärmeentwicklung sich kundgebenden gesammten [sic] Energie sein, die darüber entscheiden, in welchem Sinne die chemische Verwandschaft thätig werden kann."

Starting with Clausius' formulation of internal energy and entropy, Helmholtz wrote,

In the presentation of his general laws, Mr. Clausius uses two functions of the temperature and of a parameter he has retained, which he calls the energy U and the entropy S. Both are, however, not independent, but are rather connected through the differential equation:[2]

$$\frac{\partial S}{\partial T} = \frac{1}{T} \cdot \frac{\partial U}{\partial T}$$ (B.3)

It will be shown that both of these can be expressed through a differential quotient of the ergal,[3] which is completely defined as a function of the temperature. Thus, the thermodynamic equations no longer demand two functions of the variables, but rather only one: the ergal.[4]

The quantity he was describing was the free energy. It was the quantitative, determining value in a chemical reaction because it provided the link between total energy and entropy. Helmholtz then gave his representations of the thermodynamic equations. The first law, or *Constanz der Energie*, was:[5]

$$dQ = \frac{\partial U}{\partial T} \cdot dT + \sum_a \left\{ \left(\frac{\partial U}{\partial p_a} + P_a \right) dp_a \right\}$$ (B.4)

where $\sum_a P_a \cdot dp_a$ is the sum of freely interconvertible work energy which can be obtained from all possible processes during a differential change dp_a. For example, it could be the work required to move a differential quantity of charge, dp_{e^-}, around a circuit at a constant potential difference, $P_{voltage}$, so that the work done is $dW = P_{voltage} \times dp_{e^-}$. Helmholtz wrote the second law

$$\int \frac{dQ}{T} \cdot dT = 0$$ (B.5)

so that the process was reversible. For a system in which all parts have the same temperature, no heat could be produced at the cost of other forms of energy and the

---

[2] Helmholtz originally used ϑ to represent temperature. T is used here for continuity. Also, Helmholtz did not indicate which variable was to be held constant in his partial differentials.

[3] Clausius described the term *ergal* in *Die mechanische Wärmetheorie* (Clausius 1876, pp. 12–14). The English translation states (Clausius 1879, pp. 11–13): "...it has become needful to introduce a special name for the *negative* value of...the work performed; and Rankine proposed for this the term 'potential energy.' This name sets forth very clearly the character of the quantity; but it is somewhat long, and the author has ventured to propose in its place the term 'Ergal.'"

[4] "Hr. Clausius braucht zur Darstellung seiner allgemeinen Gesetze zwei Functionen der Temperatur und des einen von ihm beibehaltenen Parameters, welche er die Energie U und die Entropie S nennt. Beide sind aber nicht von einander unabhängig, sondern durch die Differentialgleichung: $\frac{\partial S}{\partial T} = \frac{1}{T} \cdot \frac{\partial U}{\partial T}$ miteinander verbunden. Es wird sich zeigen, dass diese beiden durch Differentialquotienten des als Function der Temperatur vollständig bestimmten Ergals dargestellt werden können, so dass die thermodynamischen Gleichungen nicht mehr zwei Functionen der Variablen, sondern nur noch eine, nämlich das Ergal erfordern."

[5] Missing from this equation is an energy conversion factor, J, whose role was to specify the conversion of heat into work. The factor has no bearing on the derivation (Partington 1964, p. 699, footnote 3).

quantity, dQ/T, which Clausius called the entropy, S, must be an explicit function of the temperature (T) and the p_a. In this case, the full differential of the entropy was:

$$\frac{1}{T} \cdot dQ = dS = \frac{\partial S}{\partial T} \cdot dT + \sum_a \left\{ \frac{\partial S}{\partial p_a} \cdot dp_a \right\} \tag{B.6}$$

Several conditions resulted from combining Eqs. B.4 and B.6. One was the mutual definition of S and U in Eq. B.3. Another was:

$$\frac{\partial S}{\partial p_a} = \frac{1}{T} \left[\frac{\partial U}{\partial p_a} + P_a \right] \tag{B.7}$$

which, after solving for P_a, allowed Helmholtz to write:

$$P_a = \frac{\partial}{p_a} [T \cdot S - U] \tag{B.8}$$

He chose:

$$A = U - T \cdot S \tag{B.9}$$

so that A,[6] like U and S, was an explicit function of T and p_a. "The functions U and S," he wrote, "which are only defined through the values of their differential quotients [Eq. B.3], each have an arbitrary additive constant. If we denote these with α and β, it follows that in the function A, an additive term of the form

$$[\alpha - \beta \cdot T] \tag{B.10}$$

will arbitrarily remain; otherwise, this function, A, is completely defined by the equation [Eq. B.9]."[7] The unknown constant of integration that Haber determined via experimental measurement was due to the mutually dependent differential definition of U and S. It would take Nernst's work—what would become the third law—to provide a theoretical solution. Inserting Eq. B.9 into Eq. B.8 yielded:

$$P_a = -\frac{\partial A}{\partial p_a} \tag{B.11}$$

---

[6] Helmholtz originally used F to represent the free energy. A is used here for continuity.

[7] "Die Functionen U und S, welche nur durch die Grössen ihrer Differentialquotienten definirt sind [Gleichung B.3], erhalten jede eine willkürliche additive Constante. Wenn wir diese mit α und β bezeichnen, folgt, dass in der Function F ein additives Glied von der Form $[\alpha - \beta \cdot T]$ willkürlich bleibt; sonst ist diese Function F durch die Gleichung [Gleichung B.9] vollständig definirt."

where A took the explicit form of a potential energy for processes at constant temperature. Differentiation of Eq. B.9 with respect to T resulted in

$$\frac{\partial A}{\partial T} = \frac{\partial U}{\partial T} - S - T \cdot \frac{\partial S}{\partial T} \tag{B.12}$$

or, after considering the definition of the U and S from Eq. B.3

$$\frac{\partial A}{\partial T} = -S \tag{B.13}$$

The definition of A from Eq. B.9 led to:

$$U = A - T \cdot \frac{\partial A}{\partial T} \tag{B.14}$$

Eq. B.14 is identical to Eq. B.1 and it can be seen that the principle of maximum work, $\Delta A = \Delta U$, is the limiting case when the temperature $\theta \to 0\,K$. Equation B.1 is applicable for both the Helmholtz free energy, A, in the Helmholtz regime with U the internal energy:

$$A \equiv U - TS \quad \to \quad A = U + T \left(\frac{\partial A}{\partial T} \right)_V \tag{B.15}$$

and for the Gibbs energy, G, in the Gibbs regime with the enthalpy $H = U + PV$:

$$G \equiv H - TS \quad \to \quad G = H + T \left(\frac{\partial G}{\partial T} \right)_P \tag{B.16}$$

Although Helmholtz discussed chemical reactions and changes of phase in his publication, he did not develop his derivation of the free energy to include changes of state involving either process. His theory was limited to simpler operations, such as the expansion or heating of an ideal gas. In these cases, as Helmholtz noted, the additive constants resulting from the integration of the expressions for energy and entropy are the same before and after the process so that they drop out of any calculation. However, a discontinuity appears both in the energy and entropy if a process involves a chemical reaction or phase change due to the heat of reaction or latent heat, respectively. In this way, the reactants (or initial phase) cease to exist along with their total energy and entropy while the products (or final phase) come into existence with their own total energy and entropy. The constants of integration are then different at the beginning and end of the process and no longer cancel out. To fully describe a chemical reaction reaching equilibrium, the constants had to be determined in some manner (Müller and Müller 2009, pp. 260–272). The difference could be measured, as Haber and others did, but the link to the heat of reaction or latent heat, that is, to thermal data, was not explicitly made until the beginning of the twentieth century by Walther Nernst.

In his 1906 publication, Nernst began with Helmholtz' solution and proceeded by introducing the basic problem: the non-equivalence of heat of reaction and work, and the resulting need to determine the constant of integration in Eq. B.2. Beginning with conventional physicochemical tools, the differential for the free energy was:[8]

$$A = U + T \frac{dA}{dT} \tag{B.17}$$

and van't Hoff's equation was written:

$$Q = RT^2 \frac{d \ln K}{dT} \tag{B.18}$$

where Q is the heat of reaction, T the temperature, and K the equilibrium constant as defined in Eq. A.18. For Q and dQ/dT Nernst set

$$Q = Q_0 + T \, \Sigma \nu \alpha + T^2 \, \Sigma \beta + \cdots \tag{B.19}$$

so that

$$\frac{dQ}{dT} = \Sigma \nu C_\nu = \Sigma \nu \alpha + 2T \, \Sigma \nu \beta + \cdots \tag{B.20}$$

where Q_0 is the heat of reaction at $T = 0\,K$, C_ν the molar heat capacities at constant volume, ν the stoichiometric coefficients, and α, β, $\cdots$ are unique constants determined by the differences in heat capacities of the reacting substances. They are not the same α and β used above by Helmholtz in Eq. B.10! Finally, after solving Eq. B.18 for $d \ln K$ and substituting Eq. B.19 for Q, integration led to

$$\ln K = -\frac{Q_0}{RT} + \frac{\Sigma \nu \alpha}{R} \ln T + \frac{\Sigma \nu \beta}{R} T + \cdots + J \tag{B.21}$$

where J is the resulting unknown constant (notice the similarity to Haber's approximation in Eq. A.23). In order to determine J, Nernst took two additional steps. First, he formulated the equations in terms of variables describing phase transitions, the values for which were available from vapor pressure experiments. Each equation, B.19–B.21, has an analogue describing the system at equilibrium between two phases—vapor/solid or vapor/liquid:

$$\lambda - RT = \lambda_0 + T(\alpha - \alpha_0) + T^2(\beta - \beta_0) + \cdots \tag{B.22}$$

$$C_\nu - c_0 = \alpha - \alpha_0 + 2T(\beta - \beta_0) + \cdots \tag{B.23}$$

$$\ln \xi = -\frac{\lambda_0}{RT} + \frac{\alpha - \alpha_0}{R} \ln T + \frac{\beta - \beta_0}{R} T + \cdots + i \tag{B.24}$$

---

[8] Nernst used neither a partial derivative nor indicated a constant volume. U was referred to as the "change in the total energy (Aenderung der gesamten Energie)."

Equation B.22 is the heat of condensation, corrected for external work. Equation B.23 is the difference between the molar heat capacities of the vapor (C_v) and solid or liquid phase (c_0). Equation B.24, the equilibrium concentration, is an integrated form of the Clausius-Clapeyron equation with i the constant of integration. It contains the vapor pressure, p, through the relationship $p = \xi RT$.

Nernst's second step was the idea that as the chemical system tends toward the absolute zero of temperature, the free energy and the heat transfer approach the same value in a special way: they converge as both their rates of change become zero. This was Nernst's formulation of the third law of thermodynamics. "It has occurred to me for some time," he wrote in 1906, "that in the case of Galvanic combination, in which only solid state bodies and very concentrated solutions are included in the equations for the chemical processes delivering [electric] current, the differences between A and U are conspicuously small [...] And so it forced the presumption that in such cases a complete coincidence of these values occurs while approaching absolute zero. Thus, the limit

$$lim\frac{dA}{dT} = lim\frac{dU}{dT}, \quad for \ T = 0 \tag{B.25}$$

would result."[9, 10]

Using van't Hoff's expression for the free energy, Nernst wrote (compare with Eq. A.19)

$$A = RT(\Sigma v ln\,\xi - ln K) \tag{B.26}$$

and proceeded to find an expression for J in terms of measurable quantities.

Setting

$$U = Q - \Sigma v(\lambda - RT) \tag{B.27}$$

in accordance with the first law and substituting Eqs. B.19 and B.22 yielded

$$U = Q_0 - \Sigma v(\lambda_o - \alpha_0 T - \beta_0 T^2) \tag{B.28}$$

with differentiation resulting in

---

[9] "Schon lange war mir...aufgefallen, daß bei galvanischen Kombinationen, bei welchen in der Gleichung des stromliefernden chemischen Prozesses nur feste Körper und sehr konzentrierte Lösungen vorkommen, die Unterschiede zwischen A und U auffällig klein sind [...] So drängte sich die Annahme auf, daß in solchen Fällen in der nächsten Nähe des absoluten Nullpunktes ein völliges Zusammenfallen beider Größen stattfindet, und es würde als Grenzgesetz $lim\frac{dA}{dT} = lim\frac{dU}{dT}$ für $T = 0$ sich ergeben."

[10] In the publication, Nernst wrote $lim\frac{dA}{dT} = lim\frac{dQ}{dT}$ for $T = 0$, using Q instead of U. In this limit the differentials of Q and U are equal, but there was no consistent differentiation.

$$\frac{dU}{dT} = \Sigma v\alpha_0 + 2T\,\Sigma v\beta_0 \tag{B.29}$$

Differentiating Eq. B.26 on the other hand and substituting Eqs. B.21 and B.24 yielded

$$-\frac{dA}{dT} = \Sigma v\alpha_0 + \Sigma v\alpha_0 \ln T + R\,(J - \Sigma vi) + 2T\,\Sigma v\beta_0 \tag{B.30}$$

Comparison of Eqs. B.29 and B.30 with Eq. B.17 and the third law (Eq. B.25) resulted in several conclusions about the system as $T \to 0\,\mathrm{K}$.

At the absolute zero of temperature, according to Eq. B.1, $A = U$. To fulfill this requirement, $-dA/dT$ cannot explode when the system's temperature approaches $0\,\mathrm{K}$. The term $\Sigma v\alpha_0 \ln T$ will do just this as $\ln T \to -\infty$ for $T \to 0\,\mathrm{K}$ and an infinite rate of change for A will not allow for $A = U$ at $0\,\mathrm{K}$. Therefore, a compensation must take place in the form of

$$\Sigma v\alpha_0 = 0 \tag{B.31}$$

With this, dU/dT in Eq. B.29 tends to zero as $T \to 0\,\mathrm{K}$. According to the third law, dA/dT must approach the same value as dU/dT. This happens if the term $R(J - \Sigma vi)$ is equal to zero so that:

$$J = \Sigma vi \tag{B.32}$$

with the i's determined from vapor pressure measurements. The calculated values for J can then be used in Eq. B.21, "with which," Nernst wrote, "the given task has been solved. The initially fully unknown integration constant, J, is reduced to a sum of integration constants, i, which are unique to each substance and can be obtained through a measurement of each substance."[11] It was not only specific values of the free energy that were calculable from the third law; Eq. B.1 allowed the determination of the entire temperature dependence (Nernst 1914).

Like Helmholtz' expression for the free energy, Nernst's third law contained profound statements about nature's behavior. It also provided a starting point, for example, for Max Planck and Albert Einstein to expand thermodynamics into the low-temperature and quantum physics of the twentieth century (Haberditzl 1960; Suhling 1972), (Bartel 1989, pp. 79–83), (Renn 2006, pp. 73–78), (Kox 2006). Nernst did not explore any of these possibilities in his publication in 1906; his comments remained limited to the application of thermochemistry.

---

[11] "…womit die gestellte Aufgabe gelöst ist. Denn die zunächst völlig unbestimmte Integrationskonstante J ist dadurch auf eine Summe von Integrationskonstanten i zurückgeführt, die jeder einzelnen Substanz eigentümlich sind und durch an jeder einzelnen Substanz auszuführende Messungen ermittelt werden können."

References

Abegg, R. and G. Bodländer. 1899. Die Elektroaffinität, ein neues Prinzip der chemischen Systematik. *Zeitschrift für anorganische Chemie* 20: 453–499.

Abegg, R. 1904a. Die Valenz und das periodische System. Versuch einer Theorie der Molekularverbindung. *Zeitschrift für anorganische Chemie* 39: 330–380.

Abegg, R. 1904b. Elektrodenvorgänge und Potentialbildung bei minimalen Ionenkonzentrationen. Bemerkungen zum Thema von Haber und Bodländer. *Zeitschrift für Elektrochemie* 10: 607–609.

Abelshauser, W., W. von Hippel, J. A. Johnson, and R. G. Stokes. 2004. *German industry and global enterprise BASF: The history of a company.* Cambridge: Cambridge University Press.

Abernathy, W. J. and J. M. Utterback. 1978. Patterns of industrial innovation. *MIT Technology Review* 80: 40–47.

Abild-Pedersen, F., J. Greeley, F. Studt, J. Rossmeisl, T. R. Munter, P. G. Moses, E. Skúlason, T. Bligaard, and J. K. Nørskov. 2007. Scaling properties of adsorption energies for hydrogen-containing molecules on transition-metal surfaces. *Physical Review Letters* 99: 016105-1–016105-4.

Adichie, C. N. 2009. *The danger of a single story; TED.* https://www.ted.com/talks/chimamanda_adichie_the_danger_of_a_single_story.

Aftalion, F. 2001. *A history of the International Chemical Industry: From the "Early Days" to 2000,* 2nd ed. Philadelphia: Chemical Heritage Press.

Akhavan, J. 2004. *The chemistry of explosives,* 2nd ed. London: Royal Society of Chemistry.

Andres, R. J., T. A. Boden, F.-M. Bréon, P. Ciais, S. Davis, D. Erickson, J. Gregg, A. Jacobson, G. Marland, J. Miller, T. Oda, J. G. J. Olivier, M. R. Raupach, P. Rayner, and K. Treanton. 2012. A synthesis of carbon dioxide emissions from fossil-fuel combustion. *Biogeosciences* 9: 1845–1871.

Arrhenius, S. 1887. Über die Dissociation der in Wasser gelösten Stoffe. *Zeitschrift für physikalische Chemie* 1, 631–648.

Atkins, P. W. 1987. *Physikalische Chemie (English: Physical chemistry).* Weinheim: VCH Verlagsgesellschaft mbH.

Auerswald, P. and L. M. Branscomb. 2003. Valleys of death and Darwinian seas: Financing the invention to innovation transition in the United States. *Journal of Technology Transfer* 28: 227–239.

Ausfelder, F. and H. Durra. 2019. 2. Roadmap des Kopernikus-Projektes "Power-to-X": Flexible Nutzung erneuerbarer Ressourcen (P2X). Technical Report, Bundesministerium für Bildung und Forschung.

© The Author(s) 2022
B. Johnson, *Making Ammonia*, https://doi.org/10.1007/978-3-030-85532-1

Austin, W. 1788. Experiments on the formation of volatile alkali, and on the affinities of the phlogisticated and light inflammable airs. *Philosophical Transactions of the Royal Society of London* 78: 379–387.

Barley, S. R. 1990. The alignment of technology and structure through roles and networks. *Administrative Science Quarterly* 35: 61–103.

Bartel, H.-G. 1989. *Walther Nernst*. Biographien hervorragender Wissenschaftler, Techniker und Mediziner, Band 90. Leipzig: BSB B.G. Teubner Verlagsgesellschaft.

Bartel, H.-G. and R. P. Huebner. 2007. *Walther Nernst: Pioneer of physics and chemistry*. Singapore: World Scientific.

Basalla, G. 1988. *The evolution of technology*. Cambridge: Cambridge University Press.

BASF. 1908a. Letter to Fritz Haber, 03 March, 1908. Archiv der Max-Planck-Gesellschaft, Abt. Va, Rep. 0005: HS 2069; BASF Akten.

BASF. 1908b. Letter to Fritz Haber, 06 March, 1908. Archiv der Max-Planck-Gesellschaft, Abt. Va, Rep. 0005: HS 2069; BASF Akten.

BASF. 1908c. Letter to Fritz Haber, 06 March, 1908. Archiv der Max-Planck-Gesellschaft, Abt. Va, Rep. 0005: HS 2055; BASF Akten.

BASF. 1910a. Internal memo, 7 January, 1910. Archiv der Max-Planck-Gesellschaft, Abt. Va, Rep. 0005: HS 2085.

BASF. 1910b. Letter to Fritz Haber, 14 February, 1910. Archiv der Max-Planck-Gesellschaft, Abt. Va, Rep. 0005: HS 2085.

BASF. 1910c. Letter to Fritz Haber, 14 March, 1910. Archiv der Max-Planck-Gesellschaft, Abt. Va, Rep. 0005: HS 2087.

BASF. 1910d. Letter to Fritz Haber, 15 January, 1910. Archiv der Max-Planck-Gesellschaft, Abt. Va, Rep. 0005: HS 2085.

BASF. 1910e. Letter to Fritz Haber, 17 February, 1910. Archiv der Max-Planck-Gesellschaft, Abt. Va, Rep. 0005: HS 2085.

BASF. 1910f. Letter to Fritz Haber, 18 March, 1910. Archiv der Max-Planck-Gesellschaft, Abt. Va, Rep. 0005: HS 2087.

BASF. 1910g. Letter to Fritz Haber, 19 October, 1910. Archiv der Max-Planck-Gesellschaft, Abt. Va, Rep. 0005: HS 2085.

BASF. 1910h. Letter to Fritz Haber, 24 January, 1910. Archiv der Max-Planck-Gesellschaft, Abt. Va, Rep. 0005: HS 2085.

BASF. 1910i. Letter to Fritz Haber, 26 March, 1910. Archiv der Max-Planck-Gesellschaft, Abt. Va, Rep. 0005: HS 2087.

BASF. 1910j. Letter to Fritz Haber, 28 February, 1910. Archiv der Max-Planck-Gesellschaft, Abt. Va, Rep. 0005: HS 2085.

BASF. 1910k. Letter to Fritz Haber, 7 February, 1910. Archiv der Max-Planck-Gesellschaft, Abt. Va, Rep. 0005: HS 2085.

Béchamp, M. A. 1854. De l'action des protosels de fer sur la nitronaphtaline et la nitrobenzine. Nouvelle méthode de formation des bases organique artificielles de Zinin. *Annales de Chimie et de Physique* 42: 186–196.

Becker, H. S. 1984. *Art worlds*. University of California Press.

Bender, M., T. Roussiere, H. Schelling, S. Schuster, and E. Schwab. 2018. Coupled production of steel and chemicals. *Chemie Ingenieur Technik* 90: 1782–1805.

Bensaude-Vincent, B. and I. Stengers. 1996. *A history of chemistry*. Cambridge: Harvard University Press.

Bernthsen, A. 1913. Die synthetische Gewinnung des Ammoniaks. *Zeitschrift für angewandte Chemie* 26: 10–16.

Bernthsen, A. 1925. *Fünzig Jahre Tätigkeit in chemischer Wissenschaft und Industrie*. Heidelberg: Verlag Chemie.

Berzelius, J. J. 1835. Årsberättelse om framstegen i fysik och kemi. *Royal Swedish Academy of Sciences*.

Blum, A., M. Jähnert, C. Lehner, and J. Renn. 2017. Translation as heuristics: Heisenberg's turn to matrix mechanics. *Studies in History and Philosophy of Modern Physics* 60: 3–22.

Bodländer, G. 1902. Betrag zur Theorie einiger technischer Reduktions- und Oxydationsprozesse. *Zeitschrift für Elektrochemie* 8: 833–843.

Bodländer, G. 1904a. Zeitgrößen der Komplexbildung, Komplexkonstanten und atomistische Dimensionen. *Zeitschrift für Elektrochemie* 10: 604–607.

Bodländer, G. and W. Eberlein. 1904b. Über einige komplexe Silbersalze. *Zeitschrift für anorganische Chemie* 39: 197–239.

Bohr, N. 1913. On the constitution of atoms and molecules. *Philosophical Magazine* 26: 1–25, 476–502, 857–875.

Bolin, B. 1970. The carbon cycle. *Scientific American* 223: 124–132.

Bonvillian, W. B. 2014. The new model innovation agencies: An overview. *Science and Public Policy* 41: 425–437.

Bosch, C. 1932. Nobel Prize Lecture. https://www.nobelprize.org/nobel_prizes/chemistry/laureates/1931/bosch-lecture.pdf.

Bosch, C. 1933. No. 6: Probleme grosstechnischer Hydrierungs-Verfahren In *Avhandlinger utgitt av Det Norske Videnskaps-Akademi I Oslo: I. Matematisk-Naturvidenskapelig Klasse*, 1–29. Oslo: A.W. Brøggers.

Bradfield, R. 1942. Liebig and the chemistry of the soil. In *Liebig and after Liebig: A century of progress in agricultural chemistry*, ed. F. R. Moulton, 48–55. Washington, D.C.: American Association for the Advancement of Science.

Branscomb, L. M. and P. E. Auerswald. 2002. Between invention and innovation. Technical Report, National Institute of Standards and Technology.

Brauer, E. 1954. Über Versuche zur katalytischen Gewinnung von Ammoniak aus den Elementen durch Katalyse. Archiv der Berlin-Brandenburgischen Akademie der Wissenschaften, NL W. Ostwald, Nr. 5311.

Brock, W. H. 1992. *The Fontana history of chemistry*. London: Fontana Press.

Bromham, L., R. Dinnage, and X. Hua. 2016. Interdisciplinary research has consistently lower funding success. *Nature* 534: 684–687.

Brooks, H. 1973. The state of the art: Technology assessment as a process. *International Social Science Journal (UNESCO, Paris)* 22: 247–256.

Brooks, H. 1994. The relationship between science and technology. *Research Policy* 23: 477–486

Browne, C. A. 1977. *A source book of agricultural chemistry*. New York: Arno Press.

Bruckner, T., I. A. Bashmakov, Y. Mulugetta, H. Chum, A. de la Vega Navarro, J. Edmonds, A. Faaij, B. Fungtammasan, A. Garg, E. Hertwich, D. Honnery, D. Infield, M. Kainuma, S. Khennas, S. Kim, H. B. Nimir, K. Riahi, N. Strachan, R. Wiser, and X. Zhang. 2014. Energy systems. In *Climate change 2014: Mitigation of climate change. Contribution of Working Group III to the fifth assessment report of the intergovernmental panel on climate change*, ed. O. Edenhofer, R. Pichs-Madruga, Y. Sokona, E. Farahani, S. Kadner, K. Seyboth, A. Adler, I. Baum, S. Brunner, P. Eickemeier, B. Kriemann, J. Savolainen, S. Schlömer, C. von Stechow, T. Zwickel, and J. C. Minx, 511–597. Cambridge: Cambridge University Press.

Bull, K., R. Hoft, and M. A. Sutton. 2011. Chapter 25: Coordinating European nitrogen policies between international conventions and intergovernmental organizations. In *The European Nitrogen Assessment: Sources, effects and policy perspectives*, ed. M. A. Sutton, C. M. Howard, J. W. Erisman, G. Billen, A. Bleeker, P. Grennfelt, H. van Grinsven, and B. Grizzetti. Cambridge: Cambridge University Press.

BUND. 2012. Nachhaltige Wissenschaft: Plädoyer für eine Wissenschaft für und mit der Gesellschaft. Technical Report 2, Bund für Umwelt und Naturschutz Deutschland.

Burchardt, L. 1975. *Wissenschaftspolitik im Wilhelminischen Deutschland: Vorgeschichte, Gründung und Aufbau der Kaiser-Wilhelm-Gesellschaft zur Förderung der Wissenschaften*. Göttingen: Vandenhoeck & Ruprecht.

Burt, R. S. 1992. *Structural holes: The social structure of competition*. Cambridge: Harvard University Press.

Burt, R. 2004. Structural holes and good ideas. *American Journal of Sociology* 110: 349–399.

Bush, V. 1945a. As we may think. *Atlantic Monthly, July*, 112–124.

Bush, V. 1945b. Science: The endless frontier. Technical Report, National Science Foundation, Washington, D.C.

Butterbach-Bahl, K., P. Gundersen, P. Ambus, J. Augustin, C. Beier, P. Boeckx, M. Dannenmann, B. S. Gimeno, A. Ibrom, R. Kiese, B. Kitzler, R. M. Rees, K. A. Smith, C. Stevens, T. Vesala, and S. Zechmeister-Boltenstern. 2011. Chapter 6: Nitrogen processes in terrestrial systems. In *The European nitrogen assessment: Sources, effects and policy perspectives*, ed. M. A. Sutton, C. M. Howard, J. W. Erisman, G. Billen, A. Bleeker, P. Grennfelt, H. van Grinsven, and B. Grizzetti. Cambridge: Cambridge University Press.

Campbell, C. T. 1994. Micro- and macro-kinetics: Their relationship in heterogeneous catalysis. *Topics in Catalysis* 1: 353–366.

Carlile, P. R. 2004. Transferring, translating, and transforming: An integrative framework for managing knowledge across boundaries. *Organization Science* 15: 499–616.

Cassebaum, H. 1982. *Carl Wilhelm Scheele*. Leipzig: B.G. Teubner.

Chandrasekhar, S. 1987. *Truth and beauty*. Chicago: The University of Chicago Press.

Charles, D. 2005. *Master mind: The rise and fall of Fritz Haber, the Nobel laureate who launched the age of chemical warfare*. New York: HarperCollins.

Chatelier, H. L. 1888. *Recherches expérimentales et théoriques sur les équilibres chimiques*. Paris: Dunod.

Chatelier, H. L. 1936. *De la Méthode dans les Sciences Expérimentales*. Paris: Dunod.

Clarke, B. and J. Foster 2009. Ecological imperialism and the global metabolic rift: Unequal exchange and the guano/nitrates trade. *International Journal of Comparative Sociology*, 311–334.

Clausius, R. 1876. *Die mechanische Wärmetheorie*, 2nd ed. Braunschweig: Vieweg.

Clausius, R. 1879. *The mechanical theory of heat*. London: Macmillan.

Coates, J. E. 1937. The Haber memorial lecture. *Journal of the Chemical Society, London* 2: 1642–1672.

Crookes, W. 1917. *The wheat problem: Based on remarks made in the presidential address to the British Association at Bristol in 1898*. London: Longmans, Green, and Co.

Csikszentmihályi, M. 1996. *Flow and the psychology of discovery and invention*. New York: HarperCollins.

Cushman, G. 2013. *Guano and the opening of the Pacific World: A global ecological history*. Cambridge: Cambridge University Press.

Dahl, S., A. Logadottir, C. J. H. Jacobsen, and J. K. Nørskov. 2001. Electronic factors in catalysis: the volcano curve and the effect of promotion in catalytic ammonia synthesis. *Applied Catalysis A* 222: 19–29.

Danneel, H. 1904. Zeitgrößen der Komplexbildung, Komplexkonstanten und atomistische Dimensionen. *Zeitschrift für Elektrochemie* 10: 609–610.

Davy, H. 1813. *Elements of agricultural chemistry*. London: Longman, Hurst, Rees, Orme, and Brown.

Delwiche, C. C. 1970. The nitrogen cycle. *Scientific American* 223: 136–146.

de Santillana, J. 1968. The role of art in the scientific renaissance. In *The rise of modern science: Internal or external factors*, ed. G. Basalla, 76–82. Lexington: D.C. Heath and Company.

de Saussure, N. T. 1804. *Recherches chimiques sur la végétation*. Paris: Chez la Ve. Nyon.

Dieterici, C. 1904. Über die thermischen und kalorischen Eigenschaften des Ammoniaks. *Zeitschrift für die gesamte Kälteindustrie* 11: 21–51.

Douat, C., S. Hübner, R. Engeln, and J. Benedikt. 2016. Production of nitric/nitrous oxide by an atmospheric pressure plasma jet. *Plasma Sources Science and Technology* 25: 025027–025038.

Durand, P., L. Breuer, P. J. Johnes, G. Billen, A. Butturini, G. Pinay, H. van Grinsven, J. Garnier, M. Rivett, D. S. Reay, C. Curtis, J. Siemens, S. Maberly, Ø. Kaste, C. Humborg, R. Loeb, J. de Klein, J. Hejzlar, N. Skoulikidis, P. Kortelainen, A. Lepistö, and R. Wright. 2011. Chapter 7: Nitrogen processes in aquatic ecosystems. In *The European nitrogen assessment: Sources, effects and policy perspectives*, ed. M. A. Sutton, C. M. Howard, J. W. Erisman, G. Billen, A. Bleeker, P. Grennfelt, H. van Grinsven, and B. Grizzetti. Cambridge: Cambridge University Press.

Eady, R. E. 1992. The dinitrogen-fixing bacteria. *The Prokaryotes*, vol. 1. New York: Springer.

Edelstein, S. 1961. Sir William Henry Perkin. In *Great chemists*, ed. E. Farbe. New York, London: Interscience Publishers.

Einstein, A. 1905. Die von der molekularkinetischen Theorie der Wärme geforderten Bewegung von in ruhenden Flüssigkeiten suspendierten Teilchen. *Annalen der Physik* 322: 549–560.

Emmett, P. H. and S. Brunauer. 1933. The adsorption of nitrogen by iron synthetic ammonia catalysts. *Journal of the American Chemical Society* 55: 1738–1739.

Engler, C., H. Bunte, F. Haber, and Klein. 1909. Memorandum, signed by the authors. Archiv der Max-Planck-Gesellschaft, Abt. Va, Rep. 0005: HS 1579.

Erisman, J. W., P. Domburg, B. de Haan, W. de Vries, J. Kros, and K. Sanders. 2005. The Dutch nitrogen cascade in the European perspective. *Science in China Series C: Life Sciences* 48: 827–842.

Erisman, J. W., M. A. Sutton, J. Galloway, Z. Klimont, and W. Winiwarter. 2008. How a century of ammonia synthesis changed the world. *Nature Geoscience* 1: 636–639.

Ertl, G. 1980. Surface science and catalysis-studies on the mechanism of ammonia synthesis: The P. H. Emmett award address. *Catalysis Reviews Science and Engineering* 21: 201–223.

Ertl, G. 1990. Elementary steps in heterogeneous catalysis. *Angewandte Chemie International Edition* 29: 1219–1227.

Ertl, G. 2008. Nobel prize lecture. *Angewandte Chemie International Edition* 47: 3524–3535.

Ertl, G. 2012. The arduous way to the Haber-Bosch process. *Zeitschrift für anorganische und allgemeine Chemie* 638: 487–489.

Ertl, G. 2015. Walther Nernst and the development of physical chemistry. *Angewandte Chemie International Edition* 54: 5828–5835.

Ertl, G. 2018. Private communication. Prof. Gerhard Ertl, Fritz Haber Institute of the Max Planck Society.

Ertl, G. and J. Soentgen. 2015. *N: Stickstoff–ein Element schreibt Weltgeschichte*. München: oekom Verlag.

Ertl, G., M. Huber, S. B. Lee, and N. Thiele. 1978. Formation and decomposition of nitrides on iron surfaces. *Zeitschrift für Naturforschung* 34: 30–39.

Ertl, G., M. Huber, S. B. Lee, Z. Paál, and M. Weiss. 1981. Interactions of nitrogen and hydrogen on iron surfaces. *Applications of Surface Science* 8: 373–386.

Ertl, G., D. Prigge, R. Schloegl, and M. Weiss. 1983. Characterization of ammonia synthesis catalysts. *Journal of Catalysis* 79: 359–377.

Eucken, W. 1921. *Die Stickstoffversorgung der Welt: Eine Volkswirtschaftliche Untersuchung*. Stuttgart und Berlin: Deutsche Verlags-Anstalt.

Fangerau, H. 2010. *Spinning the scientific web: Jaques Loeb (1859–1924) und sein Programm einer internationalen biomedizinischen Grundlagenforschung*. Berlin: Akademie Verlag.

Faraday, M. 1825. On new compounds of carbon and hydrogen, and on certain other products obtained during the decomposition of oil by heat. *Philosophical Transactions* 115: 440–466.

Farber, E. ed. 1966. *Milestones of modern chemistry: Original reports of the discoveries*. New York: Basic Books.

Farbwerke Hoechst, A. 1964. *Wilhelm Ostwald und die Stickstoffgewinnung aus der Luft*, vol. 5. Dokumente aus Hoechster Archiven, Beitrge zur Geschichte der Chemie, Hrg. Farbwerke Hoechst Aktiongesellschaft.

Farbwerke Hoechst AG. 1966. *Griesheimer Versuche zur Stickstoffgewinnung aus der Luft*, vol. 18. Dokumente aus Hoechster Archiven, Beitrge zur Geschichte der Chemie, Hrg. Farbwerke Hoechst Aktiongesellschaft.

Feynman, R. 1979. Photons: Corpuscles of light, 1979 Douglas Robb memorial lecture. Available at the Vega Science Trust (www.vega.org.uk).

Feynman, R. 1983. Richard Feynman on "why questions". 1983 BBC Interview.

Finck, A. 2003. Die Entwicklung der Mineraldüngung von Sprengel/Liebig bis heute und ihre globale Bedeutung. In *Die Entwicklung der Mineraldüngung und Carl Sprengel*, ed. M. Frielinghaus and C. Dalchow. Münchberg: Leibniz-Zentrum für Agrarlandschafts- und Landnuztungsforschung.

Fleck, L. 1979. *Genesis and development of a scientific fact*. Chicago: The University of Chicago Press.

Fleck, L. 1980. *Entstehung und Entwicklung einer wissenschaftlichen Tatsache: Einführung in die Lehre vom Denkstil und Denkkollektiv*. Frankfurt am Main: Suhrkamp Verlag.

Fligstein, N. 2001. Social skill and the theory of fields. *Sociological Theory*, 105–125.

Fortus, D. 2018. Private communication. Prof. David Fortus, The Weizmann Institute of Science.

Fraas, K. N. 1848. *Historisch-Enzyklopädischer Grundriß der Landwirthschaftslehre*. Stuttgart: Verlag der Franckh'schen Buchhandlung.

Frank, A. 1903. Die Nutzbarmachung des freien Stickstoffs der Luft für Landwirtschaft und Industrie. *Zeitschrift für angewandte Chemie* 16: 536–539.

Franz, C. 2017. Innovation for health: Success factors for the research-based pharmaceutical industry. In *Evolving business models: How CEOs transform traditional companies*, ed. C. Franz, T. Bieger, and A. Herrmann, 93–112. Cham: Springer.

Fred, E. B., I. L. Baldwin, and E. McCoy. 1932. *Root nodule bacteria and leguminous plants*. Madison: University of Wisconsin Press.

Friedrich, B. 2016. How did the tree of knowledge get its blossom? The rise of physical and theoretical chemistry, with an eye on Berlin and Leipzig. *Angewandte Chemie International Edition* 55: 5378–5392.

Friedrich, B. 2019. Fritz Haber at one hundred fifty: Evolving views of and on a german Jewish patriot. *Bunsen Magazin* 21: 130–144.

Friedrich, B. and J. James. 2017a. From Berlin-Dahlem to the fronts of World War I: The role of Fritz Haber and his Kaiser WIlhelm Institute in German chemical warfare. In *One hundred years of chemical warfare: Research, deployment and consequences*, ed. B. Friedrich, D. Hoffmann, J. Renn, F. Schmaltz, and M. Wolf, 25–44. Cham: Springer.

Friedrich, B. and D. Hoffmann. 2017b. Clara Immerwahr: A life in the shadow of Fritz Haber. In *One hundred years of chemical warfare: Research, deployment and consequences*, ed. B. Friedrich, D. Hoffmann, J. Renn, F. Schmaltz, and M. Wolf, 45–68. Cham: Springer.

Frielinghaus, M. and C. Dalchow (eds.). 2003. *Die Entwicklung der Mineraldüngung und Carl Sprengel*. Münchberg: Leibniz-Zentrum für Agrarlandschafts- und Landnuztungsforschung.

Galor, O. and D. N. Weil. 2000. Population, technology, and growth: From malthusian stagnation to the demographic transition and beyond. *The American Economic Review* 90: 806–828.

Gardner, A. 1865. The Miriam and Ira D. Wallach Division of Art, Prints and Photographs: Photography Collection, The New York Public Library; The New York Public Library Digital Collections. http://digitalcollections.nypl.org/items/ef5e5070-c3db-0134-f8ee-00505686a51c.

Geng, C., J. Li, and H. Schwarz. 2018. Ta_2^+-mediated ammonia synthesis from N_2 and H_2 at ambient temperature. *Proceedings of the National Academy of Sciences of the United States of America* 115: 11680–11687.

Gibbs, J. 1878a. On the equilibrium of heterogeneous substances. *Transactions of the Connecticut Academy of Arts and Sciences* 3: 108–248.

Gibbs, J. 1878b. On the equilibrium of heterogeneous substances. *Transactions of the Connecticut Academy of Arts and Sciences* 3: 343–524.

Girnus, W. 1987. Zu einigen Grundzügen der Herausbildung der physikalischen Chemie als Wissenschaftsdisziplin. In *Der Ursprung der modernen Wissenschaften: Studien zur Entstehung wissenschaftlicher Disziplinen*, ed. M. Guntau and H. Laitko. Berlin: Akademie-Verlag.

Globe, S. and et al. 1973. Science, technology, and innovation. Technical Report, National Science Foundation.

Goran, M. 1967. *The story of Fritz Haber*. Norman: University of Oklahoma Press.

Gorham, E. 1991. Biogeochemistry: Its origins and development. *Biogeochemistry* 13: 199–239.

Gottwald, F.-T. and W. Schmidt. 2003. Agrarqualität und Mineraldüngung. Gesichtspunkte des ökologischen Landbaus. In *Die Entwicklung der Mineraldüngung und Carl Sprengel*, ed. M. Frielinghaus and C. Dalchow. Münchberg: Leibniz-Zentrum für Agrarlandschafts- und Landnuztungsforschung.

Gräbe, C. 1920. *Geschichte der organischen Chemie*. Berlin: Springer.

Grandy, W. 1987. *Foundations of statistical mechanics*. Dordrecht: Springer.

Granovetter, M. S. 1973. The strength of weak ties. *American Journal of Sociology* 78: 1360–1380.

Graßhoff, G. 1994. The historical basis of scientific discovery. *Behavioral and Brain Sciences* 3: 545–546.

Graßhoff, G. 1998. The discovery of the urea cycle: Computer models of scientific discovery. In *Computer simulations in science and technology studies*, ed. P. Ahrweiler and N. Gilbert, 71–90. Berlin: Springer.

Graßhoff, G. 2003. Hans Krebs' and Kurt Henseleit's laboratory notebooks and their discovery of the urea cycle—reconstructed with computer models. In *Reworking the bench: Reseach notebooks in the history of science*, ed. F. H. Holmes, J. Renn, and H.-J. Rheinberger, 269–294. Dordrecht: Kluwer Academic Press.

Graßhoff, G. 2008. Gesprch mit Dr. J. Bednorz. In *Innovation–Begriffe und Thesen*, ed. G. Graßhoff and R. Schwinges, 133–147. Zürich: vdf Hochschulverlag.

Graßhoff, G. and M. May. 1995. From historical case studies to systematic methods of discovery. *Working Notes: AAAI Spring Symposium on Systematic Methods of Scientific Discovery* SS-95-03: 46–57.

Gray, M. W. 1990. From household economy to "rational agriculture": The establishment of liberal ideas in german agricultural thought. In *In search of liberal Germany: Studies in the history of German liberalism from 1789 to the present*, 25–54. New York: Berg.

Greenaway, F. 1966. *John Dalton and the atom*. London: Heinemann.

Greiner, M. T., T. E. Jones, S. Beeg, L. Zwiener, M. Scherzer, F. Girgsdies, S. Piccinin, M. Armbrüster, A. Knop-Gericke, and R. Schlögl. 2018. Free-atom-like d states in single-atom alloy catalysts. *Nature Chemistry* 10: 1008–1015.

Grossmann, H. 1974. Brunck. In *Das Buch der grossen Chemiker*, vol. 2. Berlin: Verlag Chemie.

Gruber, N. and N. Galloway. 2008. An earth-system perspective of the global nitrogen cycle. *Nature* 45: 293–296.

Güntz, M. 1898. Sur l'hydrure de baryum. *Comptes rendus hebdomadaires des séances de l'Académie des sciences* 132: 963–966.

Haber, F. 1896. *Experimentaluntersuchungen über Zersetzung und Verbrennung von Kohlenwasserstoffen*. München: R. Oldenburg.

Haber, F. 1898. *Grundriß der technischen Elektrochemie auf theoretischer Grundlage*. München: R. Oldenburg.

Haber, F. 1900a. Letter to Richard Abegg, 21 February, 1900. Archiv der Max-Planck-Gesellschaft, Abt. Va, Rep. 0005: HS 841.

Haber, F. 1900b. Letter to Wilhelm Ostwald, 29 February, 1900. Archiv der Berlin-Brandenburgischen Akademie der Wissenschaften, NL W. Ostwald, Nr. 1037.

Haber, F. 1901a. Letter to Richard Abegg, 12 February, 1901. Archiv der Max-Planck-Gesellschaft, Abt. Va, Rep. 0005: HS 842.

Haber, F. 1901b. Letter to Richard Abegg, 30 October, 1901. Archiv der Max-Planck-Gesellschaft, Abt. Va, Rep. 0005: HS 842.

Haber, F. 1901c. Letter to Richard Abegg, 7 February, 1901. Archiv der Max-Planck-Gesellschaft, Abt. Va, Rep. 0005: HS 842.

Haber, F. 1902a. Letter to Richard Abegg, 14 February, 1902. Archiv der Max-Planck-Gesellschaft, Abt. Va, Rep. 0005: HS 843.

Haber, F. 1902b. Letter to Richard Abegg, 3 March, 1902. Archiv der Max-Planck-Gesellschaft, Abt. Va, Rep. 0005: HS 843.

Haber, F. 1902c. Letter to Wilhelm Ostwald, 26 November, 1902. Archiv der Max-Planck-Gesellschaft, Abt. Va, Rep. 0005: HS 843.

Haber, F. 1903a. Über Hochschulunterricht und elektrochemische Technik in den Vereinigten Staaten. *Zeitschrift für Elektrochemie* 9: 291–303.

Haber, F. 1903b. Über Hochschulunterricht und elektrochemische Technik in den Vereinigten Staaten. *Zeitschrift für Elektrochemie* 9: 347–370.

Haber, F. 1903c. Letter to Wilhelm Ostwald, 29 July, 1903. Archiv der Berlin-Brandenburgischen Akademie der Wissenschaften, NL W. Ostwald, Nr. 1073.

Haber, F. 1903d. Über Hochschulunterricht und elektrochemische Technik in den Vereinigten Staaten. *Zeitschrift für Elektrochemie* 9: 379–406.

Haber, F. 1904a. Anhang: Zur Theorie der Reaktionsgeschwindigkeit in heterogenen Systemen. *Zeitschrift für Elektrochemie* 10: 156–157.

Haber, F. 1904b. Letter to Richard Abegg, 9 September, 1904. Archiv der Max-Planck-Gesellschaft, Abt. Va, Rep. 0005: HS 845.

Haber, F. 1904c. Letter to Richard Lorenz, 26 November, 1904. Archiv der Max-Planck-Gesellschaft, Abt. Va, Rep. 0005: HS 845.

Haber, F. 1904d. Letter to Wilhelm Ostwald, 10 February, 1904. Archiv der Max-Planck-Gesellschaft, Abt. Va, Rep. 0005: HS 845.

Haber, F. 1904e. Zeitgrößen der Komplexbildung, Komplexkonstanten und atomistische Dimensionen. *Zeitschrift für Elektrochemie* 27: 433–436.

Haber, F. 1904f. Über die kleinen Konzentrationen. *Zeitschrift für Elektrochemie* 40: 773–776.

Haber, F. 1905a. Letter to Richard Lorenz, 1905. Archiv der Max-Planck-Gesellschaft, Abt. Va, Rep. 0005: HS 846.

Haber, F. 1905b. *Thermodynamik technischer Gasreaktionen*. München: R. Oldenburg Verlag.

Haber, F. 1906. Letter to Richard Abegg, 12 June, 1906. Archiv der Max-Planck-Gesellschaft, Abt. Va, Rep. 0005: HS 847.

Haber, C. 1907a. Letter to Richard Abegg, 25 July, 1907. Archiv der Max-Planck-Gesellschaft, Abt. Va, Rep. 0005: IIS 813.

Haber, F. 1907b. Letter to Richard Abegg, 2 December, 1907. Archiv der Max-Planck-Gesellschaft, Abt. Va, Rep. 0005: HS 848.

Haber, F. 1907c. Letter to Richard Abegg, no date, but filed as 1907. Archiv der Max-Planck-Gesellschaft, Abt. Va, Rep. 0005: HS 848.

Haber, F. 1908a. Letter to BASF, 15 February, 1908. Archiv der Max-Planck-Gesellschaft, Abt. Va, Rep. 0005: HS 2069; BASF Akten.

Haber, F. 1908b. Letter to BASF, 18 January, 1908. Archiv der Max-Planck-Gesellschaft, Abt. Va, Rep. 0005: HS 2069; BASF Akten.

Haber, F. 1908c. Letter to BASF, 27 February, 1908. Archiv der Max-Planck-Gesellschaft, Abt. Va, Rep. 0005: HS 2069; BASF Akten.

Haber, F. 1908d. *Thermodynamics of technical gas reactions*. London: Longmans, Green & Co.

Haber, F. 1909a. Letter to August Bernthsen, 30 October, 1909. Archiv der Max-Planck-Gesellschaft, Abt. Va, Rep. 0005: HS 2081.

Haber, F. 1909b. Letter to BASF, 03 July, 1909. Archiv der Max-Planck-Gesellschaft, Abt. Va, Rep. 0005: HS 2080.

Haber, F. 1909c. Letter to BASF, 10 August, 1909. Archiv der Max-Planck-Gesellschaft, Abt. Va, Rep. 0005: HS 2081.

Haber, F. 1909d. Letter to Carl Engler, 18 November, 1909. Archiv der Max-Planck-Gesellschaft, Abt. Va, Rep. 0005: HS 1579.

Haber, F. 1909e. Letter to Paul Krassa, 20 December, 1909. Archiv der Max-Planck-Gesellschaft, Abt. Va, Rep. 0005: HS 850.

Haber, F. 1910a. Gewinnung von Salpetersäure aus Luft. *Zeitschrift für angewandte Chemie* 23: 684–689.

Haber, F. 1910b. Letter to BASF, 10 February, 1910. Archiv der Max-Planck-Gesellschaft, Abt. Va, Rep. 0005: HS 2085.

Haber, F. 1910c. Letter to BASF, 10 March, 1910. Archiv der Max-Planck-Gesellschaft, Abt. Va, Rep. 0005: HS 2087.

Haber, F. 1910d. Letter to BASF, 17 January, 1910. Archiv der Max-Planck-Gesellschaft, Abt. Va, Rep. 0005: HS 2085.

Haber, F. 1910e. Letter to BASF, 17 March, 1910. Archiv der Max-Planck-Gesellschaft, Abt. Va, Rep. 0005: HS 2087.

Haber, F. 1910f. Letter to BASF, 18 January, 1910. Archiv der Max-Planck-Gesellschaft, Abt. Va, Rep. 0005: HS 2085.

Haber, F. 1910g. Letter to BASF, 20 February, 1910. Archiv der Max-Planck-Gesellschaft, Abt. Va, Rep. 0005: HS 2085.

Haber, F. 1910h. Letter to BASF, 23 January, 1910. Archiv der Max-Planck-Gesellschaft, Abt. Va, Rep. 0005: HS 2085.

Haber, F. 1910i. Letter to BASF, 26 March, 1910. Archiv der Max-Planck-Gesellschaft, Abt. Va, Rep. 0005: HS 2087.

Haber, F. 1910j. Letter to BASF, 28 February, 1910. Archiv der Max-Planck-Gesellschaft, Abt. Va, Rep. 0005: HS 2085.

Haber, F. 1910k. Letter to Franz Böhm, 19 September, 1910. Archiv der Max-Planck-Gesellschaft, Abt. Va, Rep. 0005: HS 851.

Haber, F. 1910l. Über die Darstellung des Ammoniaks aus Stickstoff und Wasserstoff. *Zeitschrift für Elektrochemie* 16: 244–246.

Haber, F. 1910m. Über die Darstellung des Ammoniaks aus Stickstoff und Wasserstoff. *Chemiker Zeitung* 40: 344–346.

Haber, F. 1911. Letter to Wilhelm Ostwald, 2 February, 1911. Archiv der Max-Planck-Gesellschaft, Abt. Va, Rep. 0005: HS 852.

Haber, F. 1913. Die Vereinigung des elementaren Stickstoffs mit Sauerstoff und mit Wasserstoff. *Zeitschrift für angewandte Chemie: Wirtschaftlicher Teil* 26: 323–326.

Haber, F. 1914a. Letter to Alwin Mittasch, 13 June, 1914. Archiv der Max-Planck-Gesellschaft, Abt. Va, Rep. 0005: HS 1138.

Haber, F. 1914b. Untersuchungen über Ammoniak. Sieben Mitteilungen. I. Allgemeine Übersicht des Stoffes und der Ergebnisse. *Zeitschrift für Elektrochemie* 20: 597–604.

Haber, F. 1914c. Über die synthetische Gewinnung des Ammoniaks. *Zeitschrift für angewandte Chemie* 27: 473–477.

Haber, F. 1920. Nobel prize lecture. In English: https://www.nobelprize.org/nobel_prizs/chemistry/laureates/1918/haber-lecture.pdf, In German: https://www.nobelprize.org/nobel_prizes/chemistry/laureates/1918/haber-lecture_ty.html.

Haber, F. 1924. *Fünf Vorträge aus den Jahren 1920–1923*. Berlin: Springer.

Haber, L. F. 1958. *The chemical industry during the nineteenth century: A study of the economic aspect of applied chemistry in Europe and North America*. Oxford: Clarendon Press.

Haber, L. F. 1971. *The chemical industry 1900–1930: International growth and technological change*. Oxford: Clarendon Press.

Haber, F. and A. Moser. 1905c. Das Generatorgas- und das Kohlenelement. *Zeitschrift für Elektrochemie* 11: 593–606.

Haber, F. and G. van Oordt. 1905d. Über Bildung von Ammoniak aus den Elementen. *Zeitschrift für anorganische Chemie* 43: 111–115.

Haber, F. and L. Bruner. 1904g. Das Kohlenelement, eine Knallgaskette. *Zeitschrift für Elektrochemie* 10: 697–713.

Haber, F. and F. Richardt. 1904h. Über das Wassergasgleichgewicht in der Bunsenflamme und die chemische Bestimmung von Flammentemperaturen. *Zeitschrift für anorganische Chemie* 38: 5–64.

Haber, F. and H. Schwenke. 1904i. Über die elektrochemische Bestimmung der Angreifbarkeit des Glases. *Zeitschrift für Elektrochemie* 10: 143–156.

Haber, F. and S. Tołłoczko. 1904j. Über die Reduktion der gebundenen, festen Kohlensäure zu Kohlenstoff und über elektrochemische Veränderungen bei festen Stoffen. *Zeitschrift für anorganische Chemie* 41: 407–441.

Haber, F. and G. van Oordt. 1905b. Über die Bildung von Ammoniak aus den Elementen. *Zeitschrift für anorganische Chemie* 44: 341–378.

Haber, F. and G. van Oordt. 1905f. Über die Bildung von Ammoniak aus den Elementen. *Zeitschrift für anorganische Chemie* 47: 42–44.

Haber, F. and A. König. 1907d. Über die Stickoxydbildung im Hochspannungsbogen I. Mitteilung. *Zeitschrift für Elektrochemie* 13: 725–743.

Haber, F. and R. L. Rossignol. 1907e. Über das Ammoniakgleichgewicht. *Berichte der Deutschen Chemischen Gesellschaft* 40: 2144–2154.

Haber, F. and A. König. 1908e. Über die Stickoxydbildung im Hochspannungsbogen II. Mitteilung. *Zeitschrift für Elektrochemie* 14: 689–695.

Haber, F. and R. L. Rossignol. 1908f. Bestimmung des Ammoniakgleichgewichtes unter Druck. *Zeitschrift für Elektrochemie* 14: 181–196.

Haber, F. and R. L. Rossignol. 1908g. Die Lage des Ammoniakgleichgewichts. *Zeitschrift für Elektrochemie* 14: 513–514.

Haber, F., A. Koenig, and E. Platou. 1910. Über die Bildung von Stickoxyd im Hochspannungs-bogen. *Zeitschrift für Elektrochemie* 16: 789–796.

Haber, F. and R. L. Rossignol. 1913. Über die Technische Darstellung von Ammoniak aus den Elementen. *Zeitschrift für Elektrochemie* 19: 53–72.

Haber, F. and H. Greenwood. 1915a. Untersuchungen über Ammoniak. Sieben Mitteilungen. VII. Über die Wirkung des Urans als Katalysator bei der Synthese des Ammoniaks aus den Elementen. *Zeitschrift für Elektrochemie* 21: 241–245.

Haber, F. and A. Maschke. 1915b. Untersuchungen über Ammoniak. Sieben Mitteilungen. III. Neubestimmung des Ammoniakgleichgewichts bei gewöhnlichem Druck. *Zeitschrift für Elektrochemie* 21: 128–130.

Haber, F. and S. Tamaru. 1915c. Untersuchungen über Ammoniak. Sieben Mitteilungen. IV. Bestimmung der Bildungswärme des Ammoniaks bei hohen Temperaturen. *Zeitschrift für Elektrochemie* 21: 191–206.

Haber, F. and S. Tamaru. 1915d. Untersuchungen über Ammoniak. Sieben Mitteilungen. VI. Über die spezifische Wärme des Ammoniaks. *Zeitschrift für Elektrochemie* 21: 228–241.

Haber, F., S. Tamaru, and L. Oeholm. 1915e. Untersuchungen über Ammoniak. Sieben Mitteilun-gen. V. Über die Bildungswärme des Ammoniaks bei gewönhlicher Temperatur. *Zeitschrift für Elektrochemie* 21: 206–228.

Haber, F., S. Tamaru, and C. Ponnaz. 1915f. Untersuchungen über Ammoniak. Sieben Mitteilun-gen. II. Neubestimmung des Ammoniakgleichgewichts bei 30 Atm. Druck. *Zeitschrift für Elektrochemie* 21: 89–106.

Haberditzl, W. 1960. Walther Nernst und die Tradition der physikalischen Chemie an der Berliner Universität. *Forschen und Wirken: Festschrifte zur 150-Jahr-Feier der Humboldt-Universität zu Berlin 1810–1960* 1: 401–416.

Hager, T. 2008. *The alchemy of air: A Jewish genius, a doomed tycoon, and the scientific discovery that fed the world but fueled the rise of Hitler.* New York: Harmony Books.

Harwood, J. 2005. *Technology's dilemma: Agricultural colleges between science and practice in Germany, 1814–1934.* Oxford: Peter Lang.

Hatfield, J. L. and R. F. Follet. 2008. *Nitrogen in the environment: Sources, problems, and management.* Cambridge: Elsevier Academic Press.

Hawtof, R., S. Ghosh, E. Guarr, C. Xu, R. M. Sankran, and J. N. Renner. 2019. Catalyst-free, highly selective synthesis of ammonia from nitrogen and water by a plasma electrolytic system. *Science Advances* 5. https://doi.org/10.1126/sciadv.aat5778.

Hermann, T. A. 1979. *Deutsche Nobelpreistrger.* München: Heinz Moos Verlag.

Hellriegel, H. 1886. Welche Stickstoffquellen stehen der Pflanze zu Gebote? *Tageblatt der 59. Versammlung deutscher Naturforscher und Ärzte zu Berlin*, 290.

Hellriegel, H. and H. Wilfarth. 1888. Untersuchungen über die Stickstoffnahrung der Gramineen und Leguminosen. *Beilageheft zu der Zeitschrift des Vereins für die Rübenzucker-Industrie des Deutschen Reiches*, 234.

Hertel, O., S. Reis, C. Skjøth, A. Bleeker, R. Harrison, J. N. Cape, D. Fowler, U. Skiba, D. Simpson, T. Jickells, A. Baker, M. Kulmala, S. Gyldenkae, L. L. Sorenson, and J. W. Erisman. 2011. Chapter 9: Nitrogen processes in the atmosphere. In *The European nitrogen assessment: Sources, effects and policy perspectives*, ed. M. A. Sutton, C. M. Howard, J. W. Erisman, G. Billen, A. Bleeker, P. Grennfelt, H. van Grinsven, and B. Grizzetti. Cambridge: Cambridge University Press.

Hofmann, A. W. 1843. Chemische Untersuchung der organischen Basen im Steinkohlen-Theeröl. *Justus von Liebigs Annalen der Chemie* 47: 37–87.

Holdermann, K. 1960. *Im Banne der Chemie: Carl Bosch–Leben und Werk*. Düsseldorf: Econ Verlag.

Holmes, F. 1985. *Lavosier and the chemistry of life: An exploration of scientific creativity*. Madison: University of Wisconsin Press.

Holmes, F. 1989. *Eighteenth-century chemistry as an investigative enterprise*. Berkeley: University of California Press.

Holmes, F. 1991. *Hans Krebs: The formation of a scientific life 1900–1981*, vol. 1. Oxford: Oxford University Press.

Honcamp, F. 1930. Die Entwicklung der Landwirtschaft vor und seit dem Auftreten des Chilesalpeters. In *Hundert Jahre Chilesalpter 1830–1930*. Berlin: Komitee für Chilesalpeter.

Honkala, K., I. N. R. A. Hellman, A. Logadottir, A. Carlsson, S. Dahl, C. H. Christensen, and J. K. Nørskov. 2007. Ammonia synthesis from first-principles calculations. *Science* 307: 555–558.

Ibarra, H. and S. B. Andrews. 1993. Power, social influence, and sense making: Effects of network centrality and proximity on employee perceptions. *Administrative Science Quarterly* 38: 277–303.

Ihde, A. J. 1964. *The development of modern chemistry*. New York: Harper & Row.

Ingenhousz, J. 1798. *Über Ernährung der Pflanzen und Fruchtbarkeit des Bodens mit Einleitung von F. A. von Humboldt*. Leipzig: Schferische Buchhandlung.

Jacobsen, C. J. H., S. Dahl, B. S. Clausen, S. Bahn, A. Logadottir, and J. K. Nørskov. 2001. Catalyst design by interpolation in the periodic table: Bimetallic ammonia synthesis catalysts. *Journal of the American Chemical Society* 123: 8404–8405.

Jaenicke, J. 1958a. Johannes Jaenicke interview of George von Hevesy and Hermann Staudinger, 27 July, 1958. Archiv der Max-Planck-Gesellschaft, Abt. Va, Rep. 0005: HS 1505.

Jaenicke, J. 1958b. Johannes Jaenicke's interview of Max Mayer, 9 November, 1958. Archiv der Max-Planck-Gesellschaft, Abt. Va, Rep. 0005: HS 1483.

Jaenicke, J. 1959. Johannes Jaenicke's notes on Robert Le Rossignol's recollections of Fritz Haber. Archiv der Max-Planck-Gesellschaft, Abt. Va, Rep. 0005: HS 1496.

Jellinek, K. 1911. Einige Beobachtungen zur Rolle des Eisens als Katalysator bei der Ammoniaksynthese unter Druck. *Zeitschrift für anorganische Chemie* 71: 121–137.

Johnson, J. 1990. *The Kaiser's chemists: Science and modernization in imperial Germany*. Chapel Hill: The University of North Carolina Press.

Jones, P. M. 2017. *Agricultural enlightenment: Knowledge, technology and nature, 1750–1840*. Oxford: Oxford University Press.

Jost, F. 1908a. Über das Ammoniakgleichgewicht. *Zeitschrift für anorganische Chemie* 57: 414–430.

Jost, F. 1908b. Über die Lage des Ammoniakgleichgewichtes. *Zeitschrift für Elektrochemie* 14: 373–375.

Kant, H. 1983. *Alfred Nobel*. Biographien hervorragender Wissenschaftler, Techniker und Mediziner, Band 63. Leipzig: BSB B.G. Teubner Verlagsgesellschaft.

Kanter, R. M. 1988. When a thousand flowers bloom: Structural, collective, and social conditions for innovation in organization. *Research in Organizational Behavior* 10: 169–211.

Kasperson, C. J. 1978. Psychology of the scientist: XXXVII scientific creativity: A relationship with information channels. *Psychological Reports* 42: 691–694.

Keeling, C. D. 1973. Industrial production of carbon dioxide from fossil fuels and limestone. *Tellus* 25: 174–198.

Kekulé, A. 1865. Sur la constitution des substances aromatiques. *Bulletin de la Société Chimique de Paris* 3: 98–111.

Kennedy, P. 1987. *The rise and fall of the great powers: Economic change and military conflict from 1500–2000*. New York: Random House.

Kerckhoff, A. C. and K. W. Back. 1965. Sociometric patterns in hysterical contagion. *Sociometry* 28: 2–15.

Kleidon, A. 2016. *Thermodynamic foundations of the earth system*. Cambridge: Cambridge University Press.

Klein, U. 2015a. *Humboldts Preußen*. Darmstadt: WBG.

Klein, U. 2015b. A revolution that never happened. *Studies in History and Philosophy of Science, Part A* 49: 80–90.

Klein, U. 2016a. *Abgesang* on Kuhn's "revolutions". In *Shifting paradigms: Thomas S. Kuhn and the history of science*, 223–231. Berlin: MPIWG Edition Open Access.

Klein, U. 2016b. *Nützliches Wissen*. Göttingen: Wallstein Verlag.

Klein, U. 2016c. Useful 'knowledge'–useful 'science'. In *Preprint 481: The making of useful knowledge*, 39–48. Max Planck Institute for the History of Science.

Klein, U. 2020. *Technoscience in history: Prussia, 1750–1850*. Cambridge: MIT Press.

Kline, S. J. and N. Rosenberg. 1986. An overview of innovation. In *The positive sum strategy: Harnessing technology for economic growth*, ed. R. Landau and N. Rosenberg, 275–305. Washington, DC: National Academy Press.

König, A. 1954. Adolf König's recollections of Fritz Haber. Archiv der Max-Planck-Gesellschaft, Abt. Va, Rep. 0005: HS 1468.

Königsberg, L. 1903. *Hermann von Helmholtz*, vol. 2, 5828–5835. Braunschweig: Vieweg.

Kononova, M. M. 1966. *Soil organic matter: Its nature, its role in soil formation and in soil fertility*. Oxford: Pergamon Press.

Kox, A. J. 2006. Confusion and clarification: Albert Einstein and Walther Nernst's Heat Theorem, 1911–1916. *Studies in History and Philosophy of Modern Physics* 37: 101–114.

Koyré, A. 1968. The significance of the newtonian analysis. In *The rise of modern science: Internal or external factors*, ed. G. Basalla, 97–104. Lexington: D.C. Heath and Company.

Krassa, P. 1955. Paul Krassa's recollections of Fritz Haber. Archiv der Max-Planck-Gesellschaft, Abt. Va, Rep. 0005: HS 1470.

Krassa, P. 1966. Zur Geschichte der Ammoniaksynthese. *Chemiker-Zeitung* 90: 104–106.

Kuhn, T. S. 1959. Energy conservation as an example of simultaneous discovery. In *Critical problems in the history of science*, ed. M. Clagett, 321–356. Madison: The University of Wisconsin Press.

Kuhn, T. S. 1970. *The structure of scientific revolutions*. Chicago: The University of Chicago Press.

Laidler, K. J. 1993. *The world of physical chemistry*. Oxford: Oxford University Press.

Lalli, R., R. Howey, and D. Wintergrün. 2021. The dynamics of collaboration networks and the history of general relativity, 1925–1970. *Scientometrics* 122: 1129–1170.

Landecker, H. 2008. *Culturing life: How cells became technologies*. Cambridge: Harvard University Press.

Langrish, J. 2017. Physics or biology as models for the study of innovation. In *Critical studies of innovation*, ed. B. Godin and D. Vinck, 296–318. Cheltenham: Edward Elgar Publishing.

Laubichler, M. and J. Renn. 2015. A conceptual framework for integrating regulatory networks and niche construction. *Journal of Experimental Zoology Part B: Molecular and Developmental Evolution* 324: 1–13.

Lawes, J. B., J. H. Gilbert, and E. Pugh. 1861. On the sources of the nitrogen of vegetation; with special reference to the question whether plants assimilate free or uncombined nitrogen. *Philosophical Transactions of the Royal Society of London* 151: 431–577.

Leigh, G. J. 2004. *The world's greatest fix*. Oxford: Oxford University Press.

Lindström, B. and L. J. Pettersson. 2003. A brief history of catalysis. *Cattech* 7: 130–138.

Lunge, G. 1916. *Handbuch der Schwefelsäurefabrikation und ihrer Nebenzweige*. Braunschweig: Vieweg.

Mackinder, H. J. 1904. The geographical pivot of history. *The Geographical Journal* 23: 421–437.

Malanima, P. 2009. *Pre-modern european economy: One thousand years (10th–19th centuries)*. Leiden: Brill.

Marsch, U. 2000. Transferring strategy and structure: The German chemical industry as an exemplar for Great Britain. In *The German chemical industry in the twentieth century Vol. 8: Chemists and chemistry*, ed. J. E. Lesch. Dordrecht: Springer Science+Business Media.

Mathew, W. M. 1862. The imperialism of free trade, Peru 1820–1870. *Philosophical Transactions of the Royal Society of London* 151: 431–577.

Mazoyer, M. and L. Roudart. 2006. *A history of world agriculture (english translation)*. New York: Monthly Review Press.

McKay, H. L., S. J. Jenkins, and D. J. Wales. 2015. Theory of the $NH_x \pm H$ reactions on Fe {211}. *National Science Review* 2: 140–149.

McKee, H. S. 1962. *Nitrogen metabolism in plants*. Oxford: Clarendon Press.

Middendorff, E., B. Apolinarski, K. Becker, P. Bornkessel, T. Brandt, S. Heißenberg, and J. Poskowsky. 2017. Die wirtschaftliche und soziale Lage der Studierenden in Deutschland 2016. Zusammenfassung zur 21. Sozialerhebung des Deutschen Studentenwerks; durchgeführt vom Deutschen Zentrum für Hochschul- und Wissenschaftsforschung. Technical Report, Bundesministerium für Bildung und Forschung, Berlin.

Mitscherlich, E. 1834a. Ueber das Benzol und die Säuren der Oel- und Talgarten. *Annalen der Pharmacie* 9: 39–48.

Mitscherlich, E. 1834b. Ueber die Aetherbildung. *Annalen der Physik und Chemie* 18: 273–282.

Mittasch, A. 1939. *Kurze Geschichte der Katalyse in Praxis und Theorie*. Berlin: Springer.

Mittasch, A. 1951. *Geschichte der Ammoniaksynthese*. Weinheim: Verlag Chemie.

Moissan, H. 1898. Préparation et propriétés de l'azoture de calcium. *Comptes rendus hebdomadaires des séances de l'Académie des sciences* 127: 497–501.

Müller, I. and W. Weiss. 2005. *Energy and entropy: A universal competition*. Berlin: Springer.

Müller, I. 2007. *A history of thermodynamics: The doctrine of energy and entropy*. Berlin: Springer.

Müller, I. and W. H. Müller. 2009. *Fundamentals of thermodynamics and applications*. Berlin: Springer.

Müller, A. and D. Rehder. 2015. Biologische Stickstofffixierung: Grundlagen, Geschichte und Bedeutung In *N: Stickstoff–ein Element schreibt Weltgeschichte*, ed. G. Ertl and J. Soentgen, 47–54. München: oekom verlag.

Murmann, J. P. and K. Frenken. 2006. Toward a systematic framework for research on dominant designs, technological innovations, and industrial change. *Research Policy* 35: 925–952.

Murray, J. 1908. *Seventy-seventh meeting of the British association for the advancement of science, Leicester, 31 July–7 August, 1907* (https://archive.org/stream/reportofbritisha08scie/reportofbritisha08scie_djvu.txt). London: Spottiswoode and Co. Ltd.

Nernst, W. 1893. *Theoretische Chemie vom Standpunkte der Avogadroschen Regel und der Thermodynamik*, 1st ed. Stuttgart: Ferdinand Enke.

Nernst, W. 1903. *Theoretische Chemie vom Standpunkte der Avogadroschen Regel und der Thermodynamik*, 4th ed. Stuttgart: Ferdinand Enke.

Nernst, W. 1906. Ueber die Berechnung chemischer Gleichgewichte aus thermischen Messungen. *Nachrichten von der Königlichen Gesellschaft der Wissenschaften zu Göttingen: Mathematisch-physikalische Klasse* 1: 1–39.

Nernst, W. 1907. Ueber das Ammoniakgleichgewicht. *Zeitschrift für Elektrochemie* 13: 521–524.

Nernst, W. 1909. Die chemische Konstante des Wasserstoffs und seine Affinität zu Halogenen. *Zeitschrift für Elektrochemie* 15: 687–691.

Nernst, W. 1910. Spezifische Wärme und chemisches Gleichgewicht des Ammoniakgases. *Zeitschrift für Elektrochemie* 16: 96–102.

Nernst, W. 1914. Thermodynamische Berechnung chemischer Affinitäten. *Berichte der Deutschen Chemischen Gesellschaft* 47: 608–635.

Nernst, W. 1921. Nobel prize lecture. https://www.nobelprize.org/nobel_prizes/chemistry/laureates/1920/nernst-lecture.html.

O'Brien, T. F. 1982. *The nitrate industry and Chile's crucial transition 1870–1891*. New York: New York University Press.

Obstfeld, D. 2005. Social networks, *Tertius Iungens* orientation, and involvement in innovation. *Administrative Science Quarterly* 50: 100–130.

Obstfeld, D. 2017. *Getting new things done*. Stanford: Stanford Business Books.

Obstfeld, D. 2019. Private communication. Prof. David Obstfeld, California State University, Fullerton.

Obstfeld, D., S. P. Borgatti, and J. Davis. 2014. Brokerage as a process: Decoupling third party action from social network structure. In *Research in the sociology of organizations:*

Contemporary perspectives on organizational social networks, ed. D. J. Brass, S. P. Borgatti, D. S. Halgin, G. Labianca, and A. Mehra, vol. 4, 135–159. Bingley: Emerald Group Publishing Limited.

Ostwald, W. 1893. *Allgemeine Chemie II*, vol. Bd. I, 881. Leipzig.

Ostwald, W. 1900. Laboratory book: Elektrischer Ofen usw. (Stickstoff Problem). Archiv der Berlin-Brandenburgischen Akademie der Wissenschaften, NL W. Ostwald, Nr. 4397-3.

Ostwald, W. 1902. Vorträge und Diskussionen von der 73. Naturforscherversammlung zu Hamburg: Über Katalyse. *Physikalische Zeitschrift* 14: 313–322.

Ostwald, W. 1903. Stickstoff. Eine Lebensfrage. Archiv der Berlin-Brandenburgischen Akademie der Wissenschaften, published in Schwbischer Merkur, Stuttgart, NL W. Ostwald, Nr. 4405–01.

Ostwald, W. 1908. *Der Werdegang einer Wissenschaft: Sieben gemeinverständliche Vorträge aus der Geschichte der Chemie*. Leipzig: Akademische Verlagsgesellschaft m.b.H.

Ostwald, W. 1927. *Lebenslinien: Zweiter Teil, Leipzig 1887–1905*. Berlin: Klasin.

Padgett, J. F. 2012a. Autocatalysis in chemistry and the origin of life. In *The emergence and organization of markets*, ed. J. F. Padgett and W. W. Powell, 33–69. Princeton: Princeton University Press.

Padgett, J. F. and C. K. Ansell. 1993. Robust action and the rise of the Medici, 1400–1434. *American Journal of Sociology* 98: 1259–1319.

Padgett, J. F. and P. D. McLean. 2006. Organizational invention and elite transformation: The birth of partnership systems in renaissance florence. *American Journal of Sociology* 111: 1463–1568.

Padgett, J. F. and W. W. Powell. 2012. *The emergence and organization of markets*. Princeton University Press.

Partington, J. 1964. *A history of chemistry*, 4th ed. London: Macmillan.

Patil, B. S., F. J. J. Peeters, G. J. von Rooij, J. A. Medrano, F. Gallucci, J. Lang, Q. Wang, and V. Hessel. 2018. Plasma assisted nitrogen oxide production from air: Using pulsed powered gliding arc reactor for a containerized plant. *AIChE Journal* 64: 526–537.

Pavitt, K. 1990. What makes basic research economically useful? *Research Policy* 20: 109–119.

Perkin, W. H. 1862. On colouring matters derived from coal tar. *Journal of the Chemical Society* 1: 230–255.

Perman, E. P. 1904. The decomposition and synthesis of ammonia. *The Chemical News and Journal of Physical Science* 90: 182.

Perman, E. P. 1905. The direct synthesis of ammonia. *Proceedings of the Royal Society of London* 76: 167–174.

Perman, E. P. and G. Atkinson. 1904a. The decomposition of ammonia by heat. *Proceedings of the Royal Society of London* 74: 110–117.

Perman, E. P. and G. Atkinson. 1904b. The decomposition of ammonia by heat. *The Chemical News and Journal of Physical Science* 90: 13–17.

Perman, E. P. and J. H. Davies. 1906. Some physical constants of ammonia: A study of the effect of change of temperature and pressure on an easily condensible gas. *Proceedings of the Royal Society of London* 78: 28–42.

Perrin, J. 1909. Mouvement brownien et réalité moléculaire. *Annales de chimie et de physique*. 8iéme série 18: 5–114.

Perry-Smith, J. E. and C. E. Shalley. 2003. The social side of creativity: A static and dynamic social network perspective. *Academy of Management Review* 28: 89–106.

Planck, M. 1906. *Vorlesungen über die Theorie der Wärmestrahlung*. Leipzig: Johann Ambrosius-Barth Verlag.

Planck, M. 1919. *Das Wesen des Lichts*, 4. Berlin: Springer.

Planck, M. 1946. Persönliche Erinnerungen aus alten Zeiten. *Die Naturwissenschaften* 8: 230–235.

Planck, M., W. Nernst, H. Rubens, and E. Warburg. 1913. Wahlvorschlag von Max Planck für Albert Einstein zum ordentlichen Mitglied der physikalisch-mathematischen Klasse der Preußischen Akademie der Wissenschaften. Archiv der Berlin-Brandenburgischen Akademie der Wissenschaften, II–III–36, Bl. 36–37.

Poobalasuntharam, I., D. C. Madden, S. J. Jenkins, and D. A. King. 2011. Hydrogenation of N over Fe {111}. *Proceedings of the National Academy of Sciences of the United States of America* 108: 925–930.

Polanyi, M. 1962. *Personal knowledge.* Chicago: The University of Chicago Press.

Popper, K. 1935. *Logik der Forschung,* 1st ed. Wien: Springer.

Purrington, R. D. 1993. *Physics in the nineteenth century.* New Brunswick: Rutgers University Press.

Quinn, J. B. 1985. Managing innovation: Controlled chaos. *Harvard Business Review* 63: 73–84.

Rafiqul, I., C. Weber, B. Lehmann, and A. Voss. 2008. Energy efficiency improvements in ammonia production—perspectives and uncertainties. *Energy* 30: 2487–2504.

Ramsay, W. and S. Young. 1884. The decomposition of ammonia by heat. *Journal of the Chemical Society, Transactions* 45: 88–93.

Reay, D. S., C. M. Howard, A. Bleeker, P. Higgins, K. Smith, H. Westhoek, T. Rood, M. R. Theobald, A. Sans-Cobeña, R. M. Rees, D. Moran, and S. Reis. 2011. Chapter 26: Societal choice and communicating the European nitrogen challenge. In *The European nitrogen assessment: Sources, effects and policy perspectives,* ed. M. A. Sutton, C. M. Howard, J. W. Erisman, G. Billen, A. Bleeker, P. Grennfelt, H. van Grinsven, and B. Grizzetti. Cambridge: Cambridge University Press.

Reichenbach, H. 1954. *The rise of scientific philosophy,* 2nd ed. University of California Press.

Reinhardt, C. 1993. Über Wissenschaft und Wirtschaft. Fritz Habers Zusammenarbeit mit der BASF 1908 bis 1911. In *Naturwissenschaft und Technik in der Geschichte: 25 Jahre Lehrstuhl für Geschichte und Technik am Historischen Institut der Universtität Stuttgart,* 286–315. Stuttgart: Verlag für Geschichte der Naturwissenschaften und der Technik.

Reinhardt, C. 1997. *Forschung in der chemischen Industrie: Die Entwicklung synthetischer Farbstoffe bei BASF und Hoechst, 1863 bis 1914.* Ph. D. thesis, Technische Universitt Bergakademie Freiberg.

Renn, J. 2006. *Auf den Schultern von Riesen und Zwergen.* Weinheim: Wiley-VCH.

Renn, J. 2012. *The globalization of knowledge in history.* Berlin: MPIWG Edition Open Access.

Renn, J., B. Johnson, and B. Steininger. 2017. Ammoniak und seine Synthese: Wie eine epochale Erfindung das Leben der Menschen und die Arbeit der Chemiker verändert. *Naturwissenschaftliche Rundschau* 10: 507–514.

Rennenberg, H., M. Dannenmann, A. Gessler, J. Kreuzwieser, J. Simon, and H. Papen. 2009. Nitrogen balance in forest soils: Nutritional limitation of plants under climate change stresses. *Plant Biology* 11: 4–23.

Reschke, T. 1978. *Berzelius und Liebig: Ihre Briefe von 1831–1845,* 219–220. Göttingen: WiSoMed.

Rocke, A. J. 1984. *Chemical atomism in the nineteenth century: From Dalton to Cannizzaro.* Columbus: Ohio State University Press.

Rogers, E. M. 1995. *Diffusion of innovations,* 4th ed. New York: Free Press.

Rosenberg, N. 1983. *Inside the Black Box: Technology and economics.* Cambridge: Cambridge University Press.

Rossignol, R. L. 1928. Zur Geschichte der Herstellung des synthetischen Ammoniaks. *Die Naturwissenschaften* 50: 1070–1071.

Rudwick, M. J. S. 1985. *The Great Devonian controversy.* Chicago: The University of Chicago Press.

Russel, S. J. E. 1966. *A history of agricultural science in Great Britain.* London: Allen & Unwin.

Rutherford, E. 1938. Forty years of physics. In *Background to modern science,* ed. J. Needham and W. Pagel. Cambridge: Cambridge University Press.

Sanderson, S. and M. Uzumeri. 1995. Managing product families: The case for the Sony Walkman. *Research Policy* 24: 761–782.

Scherer, B. 2015. Die Monster. In *Das Anthropozän,* ed. J. Renn and B. Scherer, 226–241. Berlin: MSB Matthes & Seitz.

Schlenk, W. 1934. Nachruf an Fritz Haber. *Berichte der Deutschen Chemischen Gesellschaft* 67: A20–A24.

Schloesing, T. and A. Müntz. 1877a. Sur la Nitrification par les ferments organisés. *Comptes rendus de l'Académie des sciences* 84: 301–303.

Schloesing, T. and A. Müntz. 1877b. Sur la Nitrification par les ferments organisés. *Comptes rendus de l'Académie des sciences* 85: 1018–1020.

Schlögl, R. 2003. Catalytic synthesis of ammonia—a "never-ending story?". *Angewandte Chemie International Edition* 54: 3465–3520.

Schlögl, R. 2015a. Energiewende 2.0. *Angewandte Chemie* 27: 4512–4516.

Schlögl, R. 2015b. Heterogeneous catalysis. *Angewandte Chemie International Edition* 54: 2004–2008.

Schlögl, R. 2018. Private communication. Prof. Robert Schlögl, Max Planck Institute for Chemical Energy Conversion.

Schlögl, R. 2020. Chemische Batterien mit CO_2. *Angewandte Chemie*. https://doi.org/10.1002/anie.202007397.

Schmidt, A. 1934. *Die industrielle Chemie in ihrer Bedeutung im Weltbild und Erinnerungen an ihren Aufbau.* Berlin: De Gruyter.

Scholz, H. 1987. Die Entstehung der organischen Chemie als Teildisziplin der Chemie. In *Der Ursprung der modernen Wissenschaften: Studien zur Entstehung wissenschaftlicher Disziplinen*, ed. M. Guntau and H. Laitko, 154–167. Berlin: Akademie-Verlag.

Scholz, R. W. 2001. *Environmental literacy in science and society.* Cambridge: Cambridge University Press.

Schrock, R. 2006. Reduction of dinitrogen. *Proceedings of the National Academy of Sciences* 103: 17087.

Schumacher, W. 1864. *Die Ernährung der Pflanze. Mit besonderer Berücksichtigung der Culturgewächse und der landwirthschaftlichen Praxis nach den neuesten Forschungen für Landwirthe und Pflanzenforscher.* Berlin: G. F. Otto Müller's Verlag.

Schumpeter, J. 1942. *Capitalism, socialism and democracy*, 2nd ed. New York: Harper & Brothers.

Schumpeter, J. 1952. *Theorie der Wirtschaftlichen Entwicklung*, 5th ed. Berlin: August Raabe.

Schumpeter, J. 2012. *The theory of economic development*, 16th ed. New Brunswick: Transaction Publishers.

Schütt, H. W. 1992. *Eilhard Mitscherlich: Baumeister am Fundament der Chemie.* Deutsches Museum und Oldenbourg München.

Schwarte, M. 1920. *Die Technik im Weltkriege.* Berlin: Ernst Siegfried Mittler und Sohn.

Scott, A. 1905. Proceedings of the Royal Society of London referee report on E.P. Perman's "The Direct Synthesis of Ammonia". *The Royal Society Repository GB 117, Ref. No. RR/16/360.*

Seitzinger, S., J. A. Harrison, J. K. Böhlke, A. F. Bouwman, R. Lowrance, B. Peterson, C. Tobias, and G. van Drecht. 2015. The soil N cycle: New insights and key challenges. *SOIL* 1: 235–256.

Service, R. 2018. Ammonia—a renewable fuel made from sun, air, and water—could power the globe without carbon. *Science.* https://doi.org/10.1126/science.aau7489.

Sgourev, S. V. 2013. How Paris gave rise to cubism (and Picasso): Ambiguity and fragmentation in radical innovation. *Organization Science* 24: 1601–1617.

Sgourev, S. V. 2015. Brokerage as catalysis: How Diaghilev's *Ballets Russes* escalated modernism. *Organization Studies* 36: 343–361.

Sheppard, D. 2017. Robert Le Rossignol, 1884–1976. *Notes and Records: The Royal Society Journal for the History of Science* 71: 263–296.

Sheppard, D. 2020. *Robert Le Rossignol: Engineer of the Haber Process.* Cham: Springer.

Siggelkow, N. 2007. Persuasion with case studies. *Academy of Management Journal* 50: 20–24.

Silverman, A. 1838. Henry Le Chatelier: 1850 to 1936. *Journal of Chemical Education* 14: 555–560.

Simonton, D. K. 1984. *Genius, creativity, and leadership: Historiometric inquiries.* Cambridge: Harvard University Press.

Skaggs, J. 1994. *The Great Guano Rush.* New York: St. Martin's Press.

Slotta, R. 2015. Chilesalpeter—bis 1913 unersetzlich: Bergbau, Aufbereitung und Export. In *N: Stickstoff–ein Element schreibt Weltgeschichte*, ed. G. Ertl and J. Soentgen, 101–116. München: oekom verlag.

Smil, V. 1999. Detonator of the population explosion. *Nature* 400: 415.

Smil, V. 2001. *Enriching the earth: Fritz Haber, Carl Bosch and the transformation of world food production.* Cambridge: The MIT Press.

Smith, C. ed. 1998. *The science of energy: A cultural history of energy physics in victorian Britain.* London: The Athlone Press.

Soentgen, J. and J. Cyrys. 2015. Dicke Luft. In *N: Stickstoff–ein Element schreibt Weltgeschichte,* ed. G. Ertl and J. Soentgen, 41–46. München: oekom verlag.

Somorjai, G. A. and J. Y. Park. 2009. Concepts, instruments, and model systems that enabled the rapid evolution of surface science. *Surface Science* 603: 1293–1300.

Spencer, N. D., R. C. Schoomaker, and G. A. Somorjai. 1981. Structure sensitivity in the iron single-crystal catalysed synthesis of ammonia. *Nature* 294: 643–644.

Spencer, N. D., R. C. Schoomaker, and G. A. Somorjai. 1982. Iron single crystals as ammonia synthesis catalysts: Effect of surface structure on catalyst activity. *Journal of Catalysis* 74: 129–135.

Sprengel, C. 1819. *Nachrichten über Hofwyl in Briefen nebst einem Entwurf zu landwirth-schaftlichen Lehranstalten.* Celle: Schweiger et Pick.

Sprengel, C. 1828. Über Pflanzenhumus, Humussäure und humussaure Salze. *Archiv für die gesammte Naturlehre* 8: 145–220.

Sprengel, C. 1839. *Die Lehre vom Dünger oder Beschreibung aller bei der Landwirthschaft gebräuchlicher vegetabilischer, animalischer und mineralischer Düngermaterialien, nebst Erklärung ihrer Wirkungsart,* 1st ed. Leipzig: Müller Verlag.

Sprengel, C. 1845. *Die Lehre vom Dünger oder Beschreibung aller bei der Landwirthschaft gebräuchlicher vegetabilischer, animalischer und mineralischer Düngermaterialien, nebst Erklärung ihrer Wirkungsart,* 2nd ed. Leipzig: Müller Verlag.

Stal, L. J. 2015. Nitrogen fixation in cyanobacteria. *eLS*, 1–9. Chichester: John Wiley & Sons, Ltd.

Steiber, A. and S. Alnge. 2016. *The silicon valley model: Management for entrepreneurship.* Heidelberg: Springer.

T. Steinhauser, J. James, D. H. and B. Friedrich. 2011. *Hundert Jahre an der Schnittstelle von Chemie und Physik.* Berlin: De Gruyter.

Sterner, R. W., T. Anderson, J. J. Elser, D. O. Hessen, J. M. Hood, E. McCauley, and J. Urabe. 2008. Scale-dependent carbon:nitrogen:phosphorus seston stoichiometry in marine and freshwaters. *Limnology and Oceanography* 53: 1169–1180.

Stewart, W. M., D. W. Dibb, A. E. Johnston, and T. J. Smyth. 2005. The contribution of commercial fertilizer nutrients to food production. *Agronomy Journal* 1: 1–6.

Stigler, G. J. 1982. The process and progress of economics. *Nobel Prize Lecture.*

Stoltze, P. and J. K. Nørskov. 1985. Bridging the "pressure gap" between ultrahigh-vacuum surface physics and high-pressure catalysis. *Physical Review Letters* 55: 2502–2505.

Stoltze, P. and J. K. Nørskov. 1988. An interpretation of the high-pressure kinetics of ammonia synthesis based on a microscopic model. *Journal of Catalysis* 110: 1–10.

Stoltzenberg, D. 1994. *Fritz Haber: Chemiker, Nobelpreisträger, Deutscher, Jude.* Weinheim: VCH Verlagsgesellschaft mbH.

Stoltzenberg, D. 2004. *Fritz Haber: Chemist, Nobel Laureate, German, Jew (English Version).* Philadelphia: Chemical Heritage Foundation.

Stranges, A. 2000. Germany's synthetic fuel industry, 1927–1945. In *The German chemical industry in the twentieth century Vol. 8: Chemists and chemistry,* ed. J. E. Lesch. Dordrecht: Springer Science+Business Media.

Suhling, L. 1972. Walther Nernst und der Dritte Hauptsatz der Thermodynamik. In *RETE Strukturgeschichte der Naturwissenschaften,* ed. E. Schmauderer and I. Schneider, 331–346. Stuttgart: Verlag für die Geschichte der Naturwissenschaften und Technik.

Suhling, L. 1993. Nernst und die Ammoniaksynthese nach Haber und Bosch. In *Naturwissenschaft und Technik in der Geschichte,* ed. H. Albrecht, 343–356. Stuttgart: Verlag für die Geschichte der Naturwissenschaften und Technik.

Sutton, M. A., C. M. Howard, J. W. Erisman, W. J. Bealey, G. Billen, A. Bleeker, A. F. Bouwman, P. Grennfelt, H. van Grinsven, and B. Grizzetti. 2011. Chapter 5: The challenge to integrate nitrogen science and policies: The European nitrogen assessment approach. In *The European nitrogen assessment: Sources, effects and policy perspectives,* ed. M. A. Sutton, C. M.

Howard, J. W. Erisman, G. Billen, A. Bleeker, P. Grennfelt, H. van Grinsven, and B. Grizzetti. Cambridge: Cambridge University Press.

Sutton, M. A., C. M. Howard, J. W. Erisman, G. Billen, A. Bleeker, P. Grennfelt, H. van Grinsven, and B. Grizzetti. 2011a. Chapter 1: Assessing our nitrogen inheritance. In *The European nitrogen assessment: Sources, effects and policy perspectives*, ed. M. A. Sutton, C. M. Howard, J. W. Erisman, G. Billen, A. Bleeker, P. Grennfelt, H. van Grinsven, and B. Grizzetti. Cambridge: Cambridge University Press.

Sutton, M. A., C. M. Howard, J. W. Erisman, G. Billen, A. Bleeker, P. Grennfelt, H. van Grinsven, and B. Grizzetti (eds.). 2011b. *The European nitrogen assessment: Sources, effects and policy perspectives*. Cambridge: Cambridge University Press.

Sveiby, K.-E. (2017). Unattended consequences of innovation. In *Critical studies of innovation*, ed. B. Godin and D. Vinck, 137–155. Cheltenham: Edward Elgar Publishing.

Szöllösi-Janze, M. 1998b. *Fritz Haber 1868–1934*. München: C.H. Beck.

Szöllösi-Janze, M. 1998a. Friedrich Kirchenbauer, Diener: Die berufliche Karriere von Fritz Habers Mechaniker an der Technischen Hochschule Karlsruhe. *Jahrbuch für Universitts-geschichte* 1: 233–238.

Szöllösi-Janze, M. 2000. Losing the war but gaining ground: The German chemical industry during World War I. In *The German chemical industry in the twentieth century Vol. 8: Chemists and chemistry*, ed. J. E. Lesch. Dordrecht: Springer Science+Business Media.

Szöllösi-Janze, M. 2017. The scientist as expert: Fritz Haber and German chemical warfare during the First World War and beyond. In *One hundred years of chemical warfare: Research, deployment and consequences*, ed. B. Friedrich, D. Hoffmann, J. Renn, F. Schmaltz, and M. Wolf, 11–24. Cham: Springer.

Tamaru, K. 1991. The history of the development of ammonia synthesis. In *Catalytic ammonia synthesis*, ed. J. R. Jennings, 1–18. New York: Springer Science + Business Media.

Thomson, T. 1802. *A system of chemistry in four volumes*. Edinbourgh: Bell & Bradfute and E. Balfour.

Timm, B. 1984. The ammonia synthesis and heterogeneous catalysis: A historical review. *Proceedings of the 8th International Congress of Catalysis, Berlin (West) 2–6 July, 1984* 1: 7–27.

Travis, A. S. 1993a. *The rainbow makers: The origins of the synthetic dyestuffs industry in Western Europe*. Bethlehem: Lehigh University Press.

Travis, T. 1993b. The Haber-Bosch process: Exemplar of 20th century chemical industry. *Chemistry & Industry* 15: 581–585.

Travis, A. S. 2018. *Nitrogen capture: The growth of an international industry (1900–1940)*. Cham: Springer.

van Groenigen, J. W., D. Huygens, P. Boeckx, T. Kuyper, I. M. Lubbers, T. Rütting, and P. M. Groffman. 2015. The soil N cycle: New insights and key challenges. *SOIL* 1: 235–256.

Vandermeulen, D. 2005. *Fritz Haber: L'Esprit du Temps*, vol. 1. Paris: Delcourt.

Vandermeulen, D. 2007. *Fritz Haber: Les Héros*, vol. 2. Paris: Delcourt.

Vanhaute, E., C. O'Grada, and R. Paping. 2007. The European subsistence crisis of 1845–1850: A comparative perspective. In *When the potato failed. Causes and effects of the 'last' European subsistence crisis, 1845–1850*, ed. E. Vanhaute, C. O'Grada, and R. Paping, 15–42. Turnhout: Brepols Publishers.

van't Hoff, J. H. 1887. Die Rolle des osmotischen Druckes in der Analogie zwischen Lösungen und Gasen. *Zeitschrift für physikalische Chemie* 1: 481–508.

van't Hoff, J. H. 1898. *Vorlesungen über theoretische und physikalische Chemie, Heft I: Die chemische Dynamik*. Braunschweig: Vieweg.

Vedres, B. and D. Stark. 2010. Structural folds: Generative disruption in overlapping groups. *American Journal of Sociology* 115: 1150–1190.

Vojvodic, A. and J. K. Nørskov. 2015. New design paradigm for heterogeneous catalysts. *National Science Review* 2: 140–149.

Vojvodic, A., A. J. Medford, F. Studt, F. Abild-Pedersena, T. S. Khana, T. Bligaarda, and J. K. Nørskov. 2014. Exploring the limits: A low-pressure, low-temperature Haber-Bosch process. *Chemical Physics Letters* 598: 108–112.

von Helmholtz, H. 1882. Die Thermodynamik chemischer Vorgänge. *Sitzungsberichte der Königlichen Preussischen Akademie der Wissenschaft zu Berlin* 2: 22–39.

von Helmholtz, H. 1950. Über das Verhältnis der Naturwissenschaften zur Gesamtheit der Wissenschaft (Auszug aus seiner akademischen Festrede beim Antritt des Prorektorates der Universität Heidelberg im Jahre 1862). *Physikalische Blätter* 6: 145–152.

von Hippel, E. and G. von Krogh. 2017. Identifying viable "need-solution pairs": Problem solving without problem formulation. *Organizational Science* 27: 227–221.

von Jüptner, H. 1904a. Die freie Bildungsenergie einiger technisch wichtigen Reaktionen. *Zeitschrift für anorganische Chemie* 39: 49–68.

von Jüptner, H. 1904b. Die freie Bildungsenergie einiger technisch wichtigen Reaktionen. *Zeitschrift für anorganische Chemie* 40: 61–64.

von Jüptner, H. 1904c. Zur Kenntnis der freien Bildungsenergien. *Zeitschrift für anorganische Chemie* 42: 235–249.

von Jüptner, H. 1904d. Über die Bedeutung des Koeffizienten **B** im Ausdrucke für die Änderung der freien Energie. *Zeitschrift für anorganische Chemie* 40: 65–67.

von Leitner, G. 1993. *Der Fall Clara Immerwahr: Leben für eine humane Wissenschaft*. München: C.H. Beck.

von Liebig, J. 1837. *Anleitung zur Analyse organischer Körper*. Braunschweig: Vieweg.

von Liebig, J. 1842. *Die Thier-Chemie oder die organische Chemie in ihrer Anwendung auf Physiologie und Pathologie*, 1st ed. Braunschweig: Vieweg.

von Liebig, J. 1843. *Die Chemie in ihrer Anwendung auf Agricultur und Physiologie*, 5th ed. Braunschweig: Vieweg.

von Nagel, A. 1991. *Stickstoff: Die Chemie stellt die Ernährung sicher*. Mannheim: Schriftenreihe des Unternehmensarchivs der BASF Aktiengesellschaft.

von Wensierski, H., A. Langfeld, and L. Puchert. 2015. *Bildungsziel Ingenieurin: Biographien und Studienfachorientierungen von Ingenieurstudentinnen–eine qualitative Studie*. Opladen: Verlag Barbara Budrich.

Voss, M., A. Baker, H. W. Bange, D. Conley, S. Cornell, B. Deutsch, A. Engel, R. Ganeshram, J. Garnier, A.-S. Heiskanen, T. Jickells, C. Lancelot, A. McQuatters-Gollop, J. Middelburg, D. Schiedeck, C. P. Slomp, and D. P. Conley. 2011. Chapter 8: Nitrogen processes in coastal and marine ecosystems. In *The European nitrogen assessment: Sources, effects and policy perspectives*, ed. M. A. Sutton, C. M. Howard, J. W. Erisman, G. Billen, A. Bleeker, P. Grennfelt, H. van Grinsven, and B. Grizzetti. Cambridge: Cambridge University Press.

Waksman, S. 1942. Liebig—the humus theory and the role of humus in plant nutrition. In *Liebig and after Liebig: A century of progress in agricultural chemistry*, ed. F. R. Moulton, 56–63. Washington, D.C.: American Association for the Advancement of Science.

Wang, S., V. Petzold, V. Tripkovic, J. Kleis, J. G. Howalt, E. Skúlason, E. M. Fernandez, B. Hvolbæk, G. Jones, A. Toftelund, H. Falsig, M. Björketun, F. Studt, F. Abild-Pedersen, J. Rossmeisl, J. K. Nørskov, and T. Bligaard. 2011. Universal transition state scaling relations for (de)hydration of transition metals. *Physical Chemistry Chemical Physics* 13: 20760–20765.

Weininger, E. B. and A. Lareau. 2018. Pierre Bourdieu's sociology of education: Institutional form and social inequality. In *The Oxford handbook of Pierre Bourdieu*, ed. T. Medvetz and J. J. Sallaz, vol. 4, 253–272. Oxford: Oxford University Press.

Wendel, G. 1962. *Zur Würdigung unseres nationalen wissenschaftlichen Erbes: Dargestellt am Leben und Wirken des Physikochemikers Fritz Haber (1868–1934). Informations- und Studienmaterial der Zentralen Forschungsstelle, "Der Kampf der deutschen Chemiearbeiter um die Sicherheit des Friedens gegen Militarismus und Imperialismus und für den Sieg des Sozialismus, Reihe A Lehrmaterial–Folge 8*. Leuna-Merseburg: Institut für Marxismus-Leninismus an der Technischen Hochschule für Chemie.

Wendt, G. 1950. *Carl Sprengel und die von ihm geschaffene Mineraltheorie als Fundament der neuen Pflanzenernährungslehre*. Wolfenbüttel: Ernst Fischer Verlag.

Wendt, H. 2016. *The globalization of knowledge in the Iberian Colonial World.* Berlin: MPIWG Edition Open Access.

Whetham, W. C. D. 1904. Proceedings of the Royal Society of London referee report on E.P. Perman's "The Decomposition of Ammonia by Heat". *The Royal Society Repository GB 117, Ref. No. RR/16/221.*

Whitesides, G. 2015. Reinventing chemistry. *Angewandte Chemie International Edition* 54: 3196–3209.

Wilkinson, S., T. Holdich, Amery, Hogarth, and H. J. Mackinder. 1904. The geographical pivot of history: Discussion. *The Geographical Journal* 23: 437–444.

Windisch, K. 1892. *Bestimmung des Molekulargewichts in theoretischer und praktischer Beziehung.* Berlin: Springer.

Wintergrün, D. 2019. *Netzwerkanalysen und semantische Datenmodellierung als heuristische Instrumente für die historische Forschung.* Ph. D. thesis, Friedrich-Alexander-Universitt, Erlangen-Nürnberg.

Wissemeier, A. H. 2015. Können neue, innovative Düngemitteltypen das moderne Stickstoffproblem lösen? In *N: Stickstoff–ein Element schreibt Weltgeschichte,* ed. G. Ertl and J. Soentgen, 205–216. München: oekom verlag.

Wrigley, E. A. 2004. *Poverty, progress and population.* Cambridge: Cambridge University Press.

Wrigley, E. A. 2016. *The path to sustained growth: England's transition from an organic economy to an industrial revolution.* Cambridge: Cambridge University Press.

Yeang, C.-P. 2014. The Maxwellians: The reception and further development of Maxwell's electromagnetic theory. In *James Clerk Maxwell: Perspectives on his Life and Work,* ed. R. Flood, M. McCartney, and A. Whitaker. Oxford: Oxford University Press.

Youmans, E. L. 1856. *Youmans' Atlas of Chemistry.* New York: D. Appleton & Company.

Zemansky, M. and R. Dittman. 1997. *Heat and thermodynamics,* 7th ed. New York: The McGraw-Hill.

Ziman, J. 2000. Evolutionary models for technological change. In *Technological innovation as an evolutionary process,* ed. J. Ziman, 3–12. Cambridge: Cambridge University Press.

Zinin, N. 1842. Beschreibung einiger neuer organischer Basen, dargestellt durch die Einwirkung des Schwefelwasserstoffes auf Verbindungen der Kohlenwasserstoffe mit Untersalpetersäure. *Journal für praktische Chemie* 27: 140–153.

Zott, R. 2002. *Gelehrte im Für und Wider: Briefwechsel zwischen Adolf v. Baeyer und Wilhelm Ostwald (mit Briefen von und an Victor Meyer) sowie Briefwechsel zwischen Wilhelm Ostwald und Richard Abegg (mit Briefen oder Briefausschnitten von Fritz Haber und Clara Immerwahr sowie an Svante Arrhenius).* Münster: LIT Verlag.

Subject Index

© The Author(s) 2022
B. Johnson, *Making Ammonia*, https://doi.org/10.1007/978-3-030-85532-1

Places and People

© The Author(s) 2022
B. Johnson, *Making Ammonia*, https://doi.org/10.1007/978-3-030-85532-1